Consc

The Hard Problem Solved

By

Stephen Hawley Martin

WWW.OAKLEAPRESS.COM

Table of Contents

Author's Note

What scientists today call the "Hard Problem" is coming up with a theory that explains how the brain creates consciousness. For the past hundred years or more, ever since Scientific Materialism came to dominate the field of science, scientists have been trying to figure out how matter is able to create consciousness. This is because the undeniable implication of Scientific Materialism from a philosophical standpoint is "physicalism," the metaphysical thesis that "everything is physical," that there is "nothing over and above" the physical, or that everything supervenes on the physical. Therefore, mind or consciousness, the sense each of us has of "I am," must be produced solely by the brain, which according to Materialism is comprised of unthinking matter.

Before this subject is addressed, and an answer is given that's difficult if not impossible to disprove or refute, let me say that it is my sincere hope that this book will be read only by individuals who possess completely open minds. If you are convinced to your core that any of the statements listed below is an irrefutable, established fact, and no amount of evidence to the contrary will convince you otherwise,

I respectfully request that you please close this book now and go on to something else. The information herein is likely to upset you, and it has become clear to me that close-minded people tend to write one-star reviews of books that contain ideas with which they disagree, even though I suspect all they actually have read is the table of contents.

Do not read past this page if you truly believe that:

- There is an anthropomorphic God that created the physical universe and exists outside of it.
- Mainstream Christian doctrine and everything stated in the Bible is literally true.
- Nothing in the universe exists except material substance.
- Life came about by chance.
- Evolution works solely through random, chance mutations and the mechanism of "survival of the fittest" or "natural selection."
- Consciousness and intelligence did not exist until evolution produced a brain.
- Mind cannot produce matter.

According to the 19th century German philosopher Arthur Schopenhauer [1788-1860], "All truth passes through three stages: First, it is ridiculed. Second, it is violently opposed. Third, it is accepted as self-evident." At least two 21st century studies confirm this—one by Drew Westen, a professor in the Departments of Psychology and Psychiatry at Emory University, and the other by Jonathan Haidt, a psychology professor at the University of Virginia. Both strongly indicate that the majority of people immediately reject ideas that do not resonate with what they believe to be true. These studies show that just about all of us do so in a spontaneous way without first thinking through what has been written or said. We then use our considerable powers of logic and reasoning to come up with arguments to refute the ideas with which we disagree. It takes real willpower and intention to be open-minded and to genuinely consider that any statement might be true that does not coincide with what we believe we know. If you think Dr. Westen's and Dr. Haidt's conclusions may be wrong, turn on CNN, MSNBC, or FNC and watch commentators from different political parties—one Republican, the other Democrat—react to exactly the same set of facts. You know they will see them totally differently. This is why if you consider yourself to be opened minded and decide to read ahead, I implore you to read this book through from start to finish while doing your best not to make any snap judgments. Then let what you have read sink in for at least 24 hours. In other words, sleep on it. After that, decide what makes sense to you and what does not. If you disagree with something written here after dispassionately considering it, please write a reasoned rebuttal stating facts to back up what you think and send the rebuttal to me because I'm interested in knowing your ideas, and I may use what you write in an upcoming book, or perhaps a new edition of this one. I'll be sure to give you credit. You can reach me via the contract form on my website:

www.shmartin.com

One more note: I'd like to offer my apologies to those who have read other books of mine. You will likely come across passages and material that will seem very similar to those you have read before. This is because they are. I have used a good deal of information from other books of mine in the process of writing this one.

Chapter One
Consciousness and the Brain

What is consciousness? Hopefully we can agree that consciousness is the inner reality we all have of experiencing input from our senses of sight, smell, touch and so forth, as well as our thoughts, intentions, and feelings. Consciousness is what causes us to think we are separate and unique beings. As the French philosopher, René Descartes [1596-1650] said, "I think therefore I am." Consciousness is also what we use when making important decisions. I say "important" because perhaps the majority of what we do and the decisions we make are unconscious, including everything from breathing to operating an automobile, to entering letters and words into a software program by punching a keyboard as I am doing now. It's not necessary for me to think consciously about the location of the letters on the keyboard because I learned to type in high school, and because I do some typing every day, it has become second nature. When we make a decision consciously, however, we hold two or more possibilities in our minds and then we choose between them.

So what is the "Hard Problem" to do with consciousness, and why is it hard?

Today, in November 2019 as this is being written, almost all neuroscientists believe that consciousness is a product of the brain. These scientists also believe that material substance, matter, is all there is. This includes the brain, and yet somehow brain activity, which they believe is strictly matter interacting with matter, produces consciousness. The hard problem they have yet to solve is how inert matter that is not conscious can produce consciousness.

In the search for an answer, a few Materialists have become what are now known as "Panpsychists." Those in this new branch of Materialism believe that electrons, subatomic particles and atoms possess some degree of consciousness and that higher forms of consciousness emerge from this lower form. This theory suggests that there is a difference in degree rather than a difference in kind between consciousness in what the majority of Materialists likely still consider to be inert matter and what we know as human and animal consciousness. Nevertheless, in this view, consciousness is still a product of the brain and remains contained within the brain.

A different view, what almost all Materialists apparently regard as a renegade view, is what might be called a "top down" theory, as opposed to Panpsychism that can be described as one that is "bottom up." The top-down view is that the universe itself is conscious and that consciousness, mind, and intelligence existed before the material universe came into being with the Big Bang.

Let me say here, parenthetically, that many Scientific Materialists have a problem with the idea that the universe had a beginning because, if had a beginning, it must have come from something. And as the song from the *Sound of Music* goes, "Nothing comes from nothing, nothing ever could." A beginning also suggests there must have been a creator that was non-physical. Since Scientific Materialists believe all that exists is material substance, in their view there cannot have been a time when material substance did not exist and something that was non-physical did exist. There is clear evidence from astronomy, however, that the universe is expanding, and there is other evidence as well that supports the concept of a beginning or Big Bang. One theory among others that Scientific Materialists have put forth to counter the idea of a beginning is that the universe has always existed and that it contracts and expands over the eons. Currently, they say, it is in an expansion mode. However, beginning or no beginning, whether or not the universe has always existed, and expands and contracts, or whether it started from a big bang does not change or affect the evidence that will be presented here for a "top down" theory. Nor would it change the conclusion that will be drawn at the end of this book. A beginning isn't essential to the solution to the Hard Problem that will be offered.

Let's get back to the idea that the universe it-self is conscious and that consciousness supports and informs the physical universe. This is what I call the "Infinite Mind Theory." It suggests that lower levels of consciousness such as that of hu-mans and animals have come from and are the product of a much higher level of consciousness, i.e., the infinitely conscious and intelligent, ubiq-uitous mind that supports and informs the uni-verse. In this view, the entire universe is comprised of one mind that has become subdivided into many, what might be thought of as "offspring minds" that operate at different and lower levels of consciousness. This would include your mind and my mind.

Materialists, including the Panpsychist branch have yet to accept this idea, which they label "dualism." My experience has been that they reject it out of hand, and ignore or dismiss compelling evidence to support it, much of which will soon be presented. I'll begin by summarizing some of the findings of the Division of Perceptual Studies at the University of Virginia. Founded in 1967 by the late Dr. Ian Stevenson [1918-2007], the Division is a university-based research group devoted exclusively to the investigation of phenomena that challenge mainstream scientific paradigms regarding the nature of the mind/brain relationship. The researchers there are particularly interested in studying phenomena related to consciousness clearly functioning beyond the confines of the physical body, as well as phenomena that are directly suggestive of post-mortem survival of consciousness.

With apologies to those who have read my book, *Life After Death, Powerful Evidence You Will Never Die*, I'm going to summarize a lecture I also summarized in that book in an attempt to convince you of this. It was recorded on video when given in India in 2011 by Bruce Greyson, M.D., then the Chester Carlson Professor of Psychiatry and Director of the Division of Perceptual Studies at the

University of Virginia (he is now a professor emeritus). The bottom line takeaway of Dr. Greyson's lecture is that brains do not actually create consciousness, despite what many scientists believe. He does say, however, that this mistaken belief is understandable since evidence does exist that the brain produces consciousness. Consider what happens when a person drinks too much or gets knocked on the head. Also, it's possible to measure electrical activity in the brain during certain kinds of mental tasks and to identify correlations between different areas of the brain and different activities. We can stimulate different parts of the brain and record what happens as a result, and we can remove parts of the brain and observe the results on behavior. This suggests that the brain is involved with thinking, perception, and memory, but according to Dr. Greyson, it does not necessarily suggest the brain causes those thoughts, perceptions, and memories. What the measurements actually show are correlations, rather than causation. The truth is that thoughts, perceptions, and memories, actually occur somewhere else and then are received and processed by the brain in a way that might be compared to a television, cell phone, or radio receiver.

Western science, Dr. Greyson pointed out, is largely reductionist. It breaks everything down to its component parts, which are much easier to study than the whole, but the component parts do not always act like the whole. The brain is composed of millions of nerve cells or neurons, but a single neuron cannot formulate a thought, cannot feel angry or cold. It appears that brains can think and feel, but brain cells cannot. No one knows how many neurons are needed in order for them to collectively formulate a thought, nor do we know how a collection of neurons can think when a single neuron cannot.

Scientists get around this by saying consciousness is an emergent property of brains, a property that emerges when a large enough mass of brain cells gets together. According to Dr. Greyson, however, saying something is an emergent property is a way of saying it is a mystery that cannot be explained. It is a fact that there is no known mechanism in the brain or anywhere else that can produce non-physical things like thoughts, memories, or perceptions. The materialistic understanding of the world fails to deal with how electrical impulses, or a chemical trigger in the brain, can produce a thought or a feeling, or for that matter, anything the mind does. Despite this, according to Dr. Greyson, most scientists continue to maintain what he called, "The nineteenth century, Materialist view that the brain in some miraculous way we do not understand produces consciousness." These scientists, he said, "Discount or ignore that consciousness in extreme circumstances can function very well without a brain."

Dr. Greyson noted that the idea the mind and the brain are separate is what most people believed until a couple of hundred years ago, but in the nineteenth century western world, beginning with the Darwinians, science began exploring the idea that the physical brain might be the source of thoughts and consciousness. Ironically, as life scientists attempted to explain consciousness in terms of Newtonian physics, scientists in a different discipline, physics, were forced to move away from Newtonian laws and develop quantum mechanics in order to explain phenomena in which consciousness—what a researcher knows or doesn't know—completely changes the results of certain experiments as will be discussed in an upcoming chapter. It is as though the right hand did not know what the left hand was up to. Incredibly, this remains how things are today.

Dr. Greyson listed a number of examples in his lecture of evidence researchers with the Division of Perceptual Studies have collected that demonstrate that consciousness can exist without a brain being involved. It is a testament to the stubbornness of Materialist scientists that even though Dr. Greyson and his colleagues have been collecting this data for fifty-two years, and many papers and books have been written and published revealing a great deal of it, most western scientists are unaware of this evidence. As a result, you will soon have a leg up on many western scientists.

The evidence falls into four categories:

1. Recovery of lost consciousness in the moments or days prior to death among people who have been unconscious for prolonged periods of time.
2. Complex consciousness ability in some people who have minimal brain tissue.
3. Complex consciousness in near-death experiences when the brain is not functioning or is functioning at a greatly diminished level.
4. Memories, particularly among young children, accurately recalling details of a past life.

Deathbed Recovery of Lost Consciousness

The unexpected return of mental clarity shortly before death by patients suffering from neurological or psychiatric disorders has been reported in western medical literature for more than 250 years. There are published cases in the medical literature of patients suffering from brain abscesses, tumors, strokes, meningitis, Alzheimer's disease, schizophrenia, and mood disorders, all of whom long before had lost the ability to think or communicate. In many of these cases, evidence from brain scans or autopsies showed their brains had deteriorated to an irreversible degree, and yet in all of them, mental clarity returned in the last minutes, hours, and sometimes days before the patients' deaths. The Division of Perceptual Studies has identified 83 cases in western medical literature and has collected additional unpublished contemporary accounts wherein patients recovered complete consciousness just before death. It appears as though the damaged brain released its grip on a patient's mind and clarity returned as a result.

In 1844, a German psychiatrist named Julius reported that this occurred in 13 percent of patients who had died in his institution. In a recent investigation of end of life experiences in the United Kingdom, 70 percent of caregivers in nursing homes reported that they had observed patients suffering from dementia and confusion becoming completely lucid in their last hours before death. In a case Dr. Greyson himself investigated, a 42-year-old man developed a malignant brain tumor that rapidly grew in size. He quickly became bedridden, blind in one eye, unable to communicate, incoherent and bizarre in this behavior. He appeared unable to make any sense of his surroundings, and when members of his family touched him, he would slap as through being annoyed by an insect. He eventually stopped sleeping and would talk deliriously throughout the night making no sense. After several weeks of this, he suddenly appeared calm and began speaking coherently. He then slept peacefully. The following morning, he remained completely clear and talked with his wife, discussing his imminent death for the first time. He then stopped speaking and died.

There is no known physiological mechanism to explain this phenomenon. It is rare, but the fact that it happens has no explanation in terms of how the brain functions. It suggests the link between consciousness and the brain is more complex that most scientists think. It is as though the damaged brain prevents the person from communicating, but when the brain finally begins to die, consciousness is released from the degenerating brain.

Complex Consciousness Among People Who Have Minimal Brain Tissue

Another phenomenon is the presence of normal or even high intelligence in people who have very little brain tissue. There are rare but surprising cases of people who seem to function normally, with normal intelligence, and normal social function, despite having virtually no brain at all. In one case, published in 2007, a high school honor student who had been accepted for enrollment by Smith College underwent surgery after she was injured and knocked unconscious in an automobile accident. An x-ray of her head just before surgery revealed that she had no cerebral cortex at all. She had just a brainstem inside her skull. When the surgeon opened her skull to operate that is exactly what he found—a brainstem and that's all.

Neurologists tell us the brainstem relays motor and sensory signals to the cerebellum and the spinal cord and integrates heart function, breathing, wakefulness, and animal functions. They also tell us the brainstem does not have the connections to perform higher cognitive functions such as thinking, perceiving, making decisions, and so forth. According to scientific knowledge as it now stands, this college-bound honor student should not have been able to formulate a thought of any kind, let alone function at a high intellectual level.

Hers is not an isolated situation. Dr. Greyson pointed to dozens of cases of patients with hydrocephalus, wherein as much as 95 percent of a brain is incapacitated due to an excess of cerebrospinal fluid, and yet many with that level of affliction have normal and even above average intelligence.

Near Death Experiences

The near death experiences [NDEs] Dr. Greyson covered in the lecture were accounts given by people who had been clinically dead for a short time and then resuscitated or revived spontaneously. He said they typically have memories of vivid sensory imagery, and an extremely clear memory of what they experienced. They often describe what they experienced as seeming "more real" than their everyday life. All of this occurs under conditions of drastically altered brain function under which the materialist model would say is absolutely impossible. Such memories are reported by between ten and twenty percent of those who are revived from clinical death. Dr. Greyson has personally investigated almost one thousand cases.

The average age at the time of the near death in these cases was 31 years, but there was a very wide range. A young girl reported an experience she'd had at eight months old while undergoing kidney surgery. The oldest to experience near death Dr. Greyson has studied was 81 at the time of his heart attack. About one third of the NDEs occurred during surgical operations, a quarter during serious illness, and another quarter as a result of life-threatening accidents. The common features of NDEs can be categorized as changes in thinking, changes in emotional state, as well as paranormal and otherworldly features.

Changes in thinking include a sense of time being altered. Often people report that time stopped or ceased to exist. The change in thinking phenomenon also included a sudden revelation or change in understanding in which everything in the universe suddenly became crystal clear. There was a sense of the person's thoughts going much faster and being much clearer than usual. Finally, there was a life review—a panoramic memory in which the person's life seemed to flash before him or her.

Typical emotions reported included an overwhelming sense of peace and wellbeing, a sense of cosmic unity and of being one with everything, a feeling of complete joy, and a sense of being loved unconditionally.

The paranormal features included a sense of leaving the physical body, sometimes called an out of body experience [OBE], a sense of physical senses such as seeing and hearing becoming more vivid than ever before. Sometimes people report seeing colors and hearing sounds that do not exist in this life, and a sense of extrasensory perception, i.e., of knowing things beyond the normal ability of the senses, such as things that are happening at a remote location. Finally, some report having visions of the future and that they entered another, unearthly world or realm of existence.

Many report they came to a border they could not cross, a point of no return that if they crossed they would not have been able to return to life. Many also say they encountered a mystical or divine being, and some report seeing spirits and loved ones who died previously and seem to be welcoming them into another realm, or in some cases sending them back to life.

As a psychiatrist, the profound after effects of NDEs are of particular interest to Dr. Greyson. Near death survivors reliably report a consistent pattern of changes in attitudes, beliefs, and values, which do not seem to fade over time. They report overwhelmingly they are more spiritual because of their experience, that they have more compassion, a greater desire to help others, a greater appreciation for life as well as a stronger sense of meaning and purpose in life. A large majority reports they have a stronger belief that we survive death of the body and that they no longer fear death. About half report they have lost interest in material possessions, and many report they no longer have an interest in obtaining personal prestige, status, or in competition.

Dr. Greyson said that three features of NDEs suggest consciousness is not produced by the brain: 1) Enhanced mental function while the brain is incapacitated; 2) Accurate perceptions from outside the body, such as the ability to accurately tell doctors and nurses what they saw and heard going on in an operating room; and 3) encounters with deceased persons who convey accurate information no one else could have known, including in some instances of encounters with deceased persons that the NDE survivor could not have known were dead at the time.

In one case, a nine-year-old boy with meningitis had an NDE in which he saw several deceased relatives, including his sister who told him he had to return to his body. As soon as he returned from death, he told his parents—who had been at his bedside for 36 hours during his ordeal. His father became very upset because his daughter was at college in a different state and was perfectly healthy as far as her parents knew. The boy insisted that his sister had sent him back and had told him she had to remain.

The father left the hospital, promising his wife he would call their daughter as soon as he got home. When he tried to call her, he learned that the college officials had been trying to contact him and his wife all night to tell them the tragic news. Their daughter had been killed in an automobile accident around midnight.

By the way, there are a number of videos on YouTube about the work being done by the UVA's Division of Perceptual Studies, a number of which are more recent than the one of the 2011 lecture just summarized. Go to YouTube and search, "UVA

Division of Perceptual Studies." You will likely turn up several.

If you would like to see the video of Dr. Greyson's 2011 lecture, go to YouTube and search "Dr Bruce Greyson consciousness independent of the brain." A video of the lecture should come up at or near the top of the list.

Children Who Recall a Past Life

Dr. Greyson also recounted information about the Division of Perceptual Studies' research into children's memories of past lives. Researchers at the University of Virginia have been conducting these investigations since 1960 and as a result have in excess of 2500 cases in their files. I was quite familiar with this even before I saw Dr. Greyson's lecture because of research I had done for my book, *REINCARNATION: Good News for Open Minded Christians and Other Truth Seekers.* I have in fact twice interviewed one of the Perceptual Division's key researchers, the gentleman that now heads the Division, who has written two books on the Division's reincarnation research findings. His name is Jim B. Tucker, M.D., and he is also a child psychiatrist and professor in the University of Virginia's School of Medicine.

Anyone with an open mind who looks into this research will find it difficult to refute that reincarnation can and does happen. To give you a taste, I will relate a fascinating case history I also reported on in my book, *Afterlife, The Whole Truth*. This true story began on the First of May 2000.

Imagine you and your wife [or husband] are sound asleep. Your two-year-old son James is in his crib, asleep in the next room. Suddenly you are jarred awake.

You hear your son scream, "Plane on fire! Airplane crash!"

You rush into his room, and there he is on the bed, writhing the grip of horror, kicking and clawing at the covers as if he is trying to kick his way out of a coffin.

Over and over again, your child screams, "Plane on fire! Little man can't get out!"

What happened that night was not a single occurrence. Traumatic nightly scenes like it became the norm. The nightmares became even more terrifying, and James started screaming the name of the "little man" who couldn't get out of the plane. It was "James," his own name. Other words he spoke out loud included: "Jack Larsen," "Natoma" and "Corsair."

James' father, Bruce Leininger, could not think of what to do. Eventually, in attempt to find an answer to his son's troubled nights, he embarked on a research project, armed only with the names and words his son had been shouting while in a disturbed sleep.

A devote Christian, the answer Bruce found was not the one he wanted. He came to believe his son James was the reincarnation of a World War Two fighter pilot whose plane had been hit and downed by antiaircraft fire—a pilot named James Huston who had died in 1945 after his plane suffered a direct hit and crashed.

James' mother, however, was the first to suspect the truth. At the time, James was having five nightmares a week, and his mother, Andrea, was worried. At a toy shop, Andrea and James were looking at model planes.

"Look," Andrea said. "There's a bomb on the bottom of that one."

"That's not a bomb, Mommy," James said. "That's a drop tank."

The child was two years old. How could he possibly have known about the gas tank used by aircraft in World War Two to extend their range?

As the nightmares continued, Andrea asked, "Who is the 'little man'?"

"Me," he answered.

Bruce asked, "What happened to your plane?"

"It crashed on fire."

"Why did your plane crash?"

"It got shot," James said.

"Who shot your plane?"

"The Japanese!" he said.

James said he knew it was the Japanese because of "the big red sun." He was, of course, describing the Japanese symbol of the rising sun painted on their warplanes.

Andrea began to suggest reincarnation. Wouldn't that explain it? But Bruce reacted angrily. He thought there must be a rational explanation, and reincarnation was definitely not in his mind a rational explanation.

Bruce questioned his son further. "Do you remember what kind of plane the little man flew?"

"A Corsair," two-year-old James replied without hesitation. It was a word he had shouted in his dreams.

Bruce knew a Corsair was a World War Two fighter plane.

"Where did your airplane take off?" Bruce asked.

"A boat."

"What was the name of the boat?"

James replied with certainty, "The Natoma."

Bruce did some research. He was amazed to find the Natoma Bay was a World War Two aircraft carrier. Bruce rushed to his office, where he had a dictionary of American naval fighting ships. Natoma Bay had supported the U.S. Marines' invasion of Iwo Jima in 1945.

Andrea, meanwhile, had become convinced James was reincarnated. She contacted Carol Bowman, the author of a book on reincarnation and children who remember past lives. Bowman confirmed Andrea's views, saying that the common threads were there with James, including his age when the nightmares began and his remembered death.

Bruce kept investigating. He decided to see if he could find someone named Jack Larsen, a name James had shouted repeatedly during his nightmares. Bruce was successful in finding someone who fit the time period and place. It turned out Larsen's friend James Huston had died when his plane was shot in the engine and caught fire, just as had been described by two-year-old James Leininger.

Bruce also found Huston's name on the list of 18 men killed in action on the Natoma. The discovery finally made him realize his son might actually be the reincarnation of James Huston. But he kept investigating, anyway, and everything he found served to confirm that conclusion.

One day, little James unnerved his father when he said, "I knew you would be a good daddy, that's why I picked you."

"Where did you find us?" asked an incredulous Bruce.

"In Hawaii, at the pink hotel on the beach," James said, and went on to describe his parents' fifth wedding anniversary, which had taken place five weeks before Andrea had gotten pregnant. James said that was when he "chose" the couple to bring him back into the world.

Something new emerged almost every day. On a map, James pointed out the exact location where James Huston's plane went down. Asked why he called his action figures "Billy," "Leon" and "Walter," he replied, "Because that's who met me when I got to heaven."

Eventually, the family received a phone call from a veteran who had seen Huston's plane get hit. The man had kept his knowledge to himself for more than 50 years. He described seeing the aftermath of Huston's crash on the sea below.

"He took a direct hit on the nose. All I could see were pieces falling into the bay. We pulled out of the dive and headed for open sea. I saw the place where the fighter had hit. The rings were still expanding near a huge rock at the harbor entrance."

And so it was as James had said. His plane was hit in the engine and the front exploded in a ball of flames, but that was not the end of James. He returned to this reality fifty-three years later, in 1998, with his memory intact. Perhaps he had some things here on earth he wanted to do, like flying airplanes.

Chapter Two
Evidence Mind Is Nonlocal

In this chapter we will explore evidence that indicates mind—the medium of thought—permeates physical reality, but before we look at the facts, let's consider how the idea that only matter exists, i.e., how Scientific Materialism, came about.

Back in the seventeenth century, almost everyone believed in unseen worlds, what we would call heaven and hell, and although most people couldn't see them, these realms surrounded us—they were on the other side of the "veil." Cotton Mather (1663-1728), the New England Puritan minister, prolific author and pamphleteer, for example, wrote a book called, *The Wonders of the Invisible World.* But by the late seventeenth century, while almost everyone still believed in God, belief in an unseen world was no longer universal. An English philosopher, Thomas Hobbes (1588-1679), had argued that aside from God—the "First Cause" that created the material world—nothing existed that is not of the material world. The logic he used was simple. How could it, if God created everything?

"The universe," Hobbes wrote, "that is, the whole mass of things that are, is corporeal, that is to say, body, and hath the dimensions of magnitude, namely length, breadth, and depth; also every part of body is likewise body, and hath the like dimensions, and consequently every part of the universe is body, and that which is not body is no part of the universe: and because the universe is all, that which is no part of it is nothing, and consequently nowhere."

Hobbes had a big impact on the Age of Enlightenment, which was to pick up steam a century later when many thought leaders of the day, including Thomas Jefferson, for example, embraced Deism, the belief in the existence of a supreme being, but one that does not intervene in the universe. Also referred to as "The Clockmaker Theory," Deism is the term used to indicate an intellectual movement of the 17th and 18th centuries that accepted the existence of a creator on the basis of reason but rejected belief in a supernatural deity that interacts with humankind. Because of the limited knowledge they had to work with back then, what Hobbes argued made sense to the intellectuals of that day. Nobody, Hobbes included, knew about electromagnetism, gamma rays, radio and TV waves and more that aren't of the material world, but nonetheless are as real as anything constructed of matter and can be measured with instruments today. This continued to be true in 1859 when Darwin's *The Origin of Species* was published, which gave a theory of how evolution works through "natural selection." This theory made it possible to remove God from the equation completely. As you will see in an upcoming chapter, however, while the mechanism of "natural selection" no doubt plays an important role in evolution, it no longer makes sense as the sole producer of life and evolution when the information that will be revealed is taken into account.

Evidence Mind Is Ubiquitous

Let's take a look at a quantum mechanics experiment that I believe supports the assertion that the minds of humans and the Infinite Mind are connected. It has to do with how light behaves, and it's called the "Double Slit Experiment." The strange but revealing phenomenon associated with it is known as "The Participating Observer." The findings of this experiment are straightforward, and have been replicated in a number of different laboratories. No honest scientist will refute them.

Scientists have known for more than a hundred years that light can behave both as waves and as particles (photons), but until 1905 they thought light was comprised only of waves. Thomas Young (1773-1829) demonstrated in 1803 that light is waves by placing a screen with two parallel slits between a source of light—sunlight coming through a hole in a screen—and a wall. Each slit could be covered with a piece of cloth. These slits were razor thin, not as wide as the wavelength of the light. When waves of any kind pass through an opening not as wide as they are, the waves diffract. This was the case with one slit open. A fuzzy circle of light appeared on the wall.

When both slits were uncovered, alternating bands of light and darkness appeared, the center band being the brightest. Scientists call this a zebra pattern. The areas of light and dark result from what is known in wave mechanics as interference. Waves overlap and reinforce each other in some places and in others they cancel each other out. The bands of light on the wall indicated where one wave crest overlapped another crest. The dark areas showed where a crest and a trough met and canceled each other out.

In 1905, Albert Einstein published a paper that revealed light also behaves as if it consists of particles. He did so by using the photoelectric effect. When light hits the surface of a metal, it jars electrons loose from the atoms in the metal and sends them flying off as though struck by tiny billiard balls. So, Thomas Young demonstrated light is waves, and Einstein demonstrated it is particles. This is the sort of paradox that led scientists to develop quantum mechanics.

Now let's consider a double slit experiment constructed to determine what happens when those conducting the experiment observe or do not observe which slits the photons of light pass through. This time a gun is used that fires one photon at a time. I first read about this experiment years ago in an article entitled, "Faster Than What?" in the June 19, 1995 issue of *Newsweek*. It reported on a paper to be published by a well-known quantum physicist, Raymond Chiao, then teaching at the University of California at Berkeley. Just so you'll know I'm not making this up, much later, in July 2008, I reviewed the facts of this experiment as they are presented here with a guest on the radio show I hosted at the time, the noted quantum physicist Henry P. Stapp, author of, *MINDFUL UNIVERSE: Quantum Mechanics and the Participating Observer* (Springer, 2007). He indicated I understood the facts correctly.

Both slits were open and a detector was used to determine which slit a photon passed through. A record was made of where each one hit. Only one photon at a time was shot, so one would suppose there could be no interference. This was the case. The photons did not make a zebra pattern. Rather, they made marks, tiny dots, on a screen.

When the detector was turned off, however, and it was not known which slit a photon passed through, the zebra pattern appeared. In other words, without the detector making it possible for the researcher to observe which slit particles passed through, the particles behaved like waves even though they were fired one at a time.

Imagine the stir this caused among those conducting the experiment. In the *Newsweek* article reporting on this, Nobel Prize winning physicist Richard Feynman (1918-1988) was quoted as saying this is the "central mystery" quantum mechanics, that something as intangible as knowledge—in this case, which slit a photon went through—changes something as concrete as a pattern on a screen.

But how could knowledge change the behavior of particles shot from a gun? Materialist science cannot produce an explanation because, as we know, a tenet of Scientific Materialism is that the brain produces consciousness, awareness, and thought, and that means consciousness, awareness, and thought are confined within a person's skull. Since it would be ludicrous to suggest that thought enclosed inside a person's head could be capable of having an effect on photons shot from a gun, it ought to be clear to everyone that the tenet is false. Obviously, consciousness is not confined inside the skull. Yet Materialists tenaciously cling to the tenet, saying there must be two different sets of laws of physics: a small (subatomic world) set, and a macro world (the one we live in) set. Somewhere between these two worlds, the laws of physics must change.

This hypothesis doesn't explain why thought contained in someone's head should change things at the subatomic level, or anywhere else. Second, other experiments refute the contention two different laws of physics exist. One such experiment involves large (carbon 60) molecules called "buckyballs," so big they can be seen and therefore are part of the macro world. Research shows they exhibit the same quantum properties as infinitely small particles. Another is an experiment conducted in 2002 that involved crystals. It produced quantum ridges half an inch high—large enough to be measured with a conventional macro-world ruler.

What makes more sense is what William of Ockham (c. 1287–1347) is thought to have been the first to say, "The simplest explanation is the best." The researcher's ability to know—his or her consciousness—causes the waves to collapse into particles that form a pattern. In other words, when the researcher can access the knowledge, the zebra pattern does not occur. If he or she cannot access it, the zebra pattern appears. This was verified by setting up the experiment several ways. In the first, the detectors were in front of the two slits. In the second, researchers placed detectors between the screen and the two slits, i.e., after the photons had passed through them. As in the original experiment, knowing about a photon's behavior at the two slits made the zebra pattern vanish, whether or not the detectors were before or after the slits (see the accompanying graphic). But when the detectors were switched off, the zebra stripes returned.

In a third variation, a detector was placed before the slits and a mechanism erased the knowledge after the photon had passed through. The same thing happened. The zebra pattern returned. The result was the same no matter which way the experiment was set up—before the slits, after the

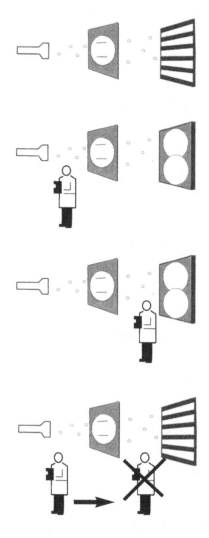

The Double Slit Experiment demonstrates the observer's mind is at one with the Infinite Mind.

slits, or before the slits and then erased. Whether or not the researcher was able to know where each photon hit determined the presence of the zebra pattern, or the lack of it. Versions of the experiment reported on in *Newsweek* were carried out at the University of Munich and at the University of Maryland. The behavior of the photons, the researchers report, "is changed by how we are going to look at them."

The question is, how? As mentioned above, the answer escapes scientists who refuse to think outside the Materialist box, but as soon as one is willing to step outside those four walls, the answer is clear. Consciousness—thought—creates reality. In the double slit experiment, it is the participating observer's consciousness that creates the resulting reality. Before the multitude of photons fired through slits is observed [seen or looked at], it exists as non-material potential in the form of waves. Observed, the waves collapse into something more solid—photons that form a pattern. What the researcher conducting the experiment thinks ought to be the result is the result—marks on a screen where the photons hit that he or she fired from the photon gun, and so the observer's mind creates the reality.

How does conventional twenty-first century science attempt to explain physical reality? Quantum physicists have developed what they call "String Theory" in an effort to explain how sub-atomic particles form out of vibrations, but they don't say what is vibrating or where the vibrations come from, according to what I have read, including *The Fabric of the Cosmos: Space, Time, and the Texture of Reality* (Alfred A. Knoph, 2004) by Rhodes Scholar and physicist, Brian Greene. I say it is mind, and that mind, which is also energy, creates matter, as in $E=MC^2$. I am about to provide evidence to support this contention, evidence that was collected by a college professor I spoke with on my radio show named Stephen Braude. At the time, Dr. Braude was a tenured professor of philosophy at the University of Maryland Baltimore County, and I had just read his book, *The Gold Leaf Lady and Other Parapsychological Investigations* (The University of Chicago Press, 2007).

The interview did not disappoint. Dr. Braude related several well documented and amazing stories of mind over matter, but perhaps the most fantastic, as well as the one that supports the contention that mind [Infinite Mind] creates matter, had to do with Katie, a woman born in Tennessee, the tenth of twelve children.

Katie is apparently a simple woman. Illiterate, at the time Dr. Braude wrote the book about her, she lived in Florida with her husband and worked as a domestic. She was also a psychic who'd had documented successes helping the police solve crimes. In one instance she was able to describe the details of a case so thoroughly and accurately, the police regarded her as a suspect until those actually responsible were apprehended. She apparently also was able to apport objects—in other words, she somehow caused them to disappear in one place and reappear in another, at least that is what Dr. Braude maintained when I spoke with him. And that wasn't not all. Seeds reportedly germinated rapidly in her cupped hands. Observers claim to have seen her bend metal, and she was both a healer and a medium or channel. Being illiterate, she cannot read or write in her native English, but she has been video taped writing quatrains in medieval French similar both in style and content to the quatrains of Nostradamus. I know some scientists are going to flip out when they read what comes next because it goes against what they see as a fixed law of physics—that matter cannot be created nor destroyed—but most amazing, perhaps, is what appeared spontaneously on her skin—on her hands, face, arms, legs, and back—apparently out of thin air. It looked like gold leaf, a thin version of the wrapping on a Hersey's Kiss. Katie could not control when this happened, but Dr. Braude and other witnesses saw the foil materialize firsthand. He even videotaped it appearing on her skin.

I just stopped typing and checked. As of this writing, footage from this video can be seen on YouTube. Go to YouTube and put "gold leaf lady Braude" in the YouTube search bar. The title of the video is "UMBC In the Loop: Stephen Braude."

Dr. Braude took the foil to be analyzed. It turned out not to be gold at all, but brass—approximately 80 percent copper and 20 percent zinc.

Dr. Braude thinks there's a reason she produces brass and not gold. Where does the brass foil that appears on Katie's skin come from? It appears that her mind creates it. In fact, as mentioned, Dr. Braude believes she produces brass rather than gold for a reason. You see, Katie has a difficult and tense relationship with her husband. Once she apported a carving set. It just appeared. And her husband—apparently nonplussed—said, "So what? It's not worth anything." Soon afterward, gold colored foil began appearing on Katie's skin. But it wasn't real gold, it was fool's gold—brass. Dr. Braude thinks this is how she gets back at her husband. Katie's mind—albeit the unconscious part—creates matter in the form of brass foil. This being the case, why should it be difficult to believe that an Infinite Mind—one infinitely more powerful than a human mind—created the material universe? The physical universe had to come from something.

One thing quantum physicists agree on is that the physical universe isn't really physical. It is energy—vibrations—in other words, waves. In the double slit experiment, the researcher's ability to know turned waves into photons, which is to say that thought or mind turned waves into things [particles]. It does not seem to me too big a leap to suggest that mind is what creates physical reality.

Whether or not the physical universe was created with a Big Bang or has always existed, to me it makes sense that on an infinitely larger scale than the double slit experiment, it is an Infinite Mind that creates physical reality.

Chapter Three
The Complexity of Physical Reality and Life

In a book to be released in Spring 2020, *The Return of the God Hypothesis: Compelling Scientific Evidence for the Existence of God,* by Stephen C. Meyer, the *New York Times* bestselling author of *Darwin's Doubt* and *Intelligent Design* will present what his publisher calls "groundbreaking scientific evidence of the existence of God, based on breakthroughs in physics, cosmology, and biology."

In his previous bestsellers, Meyer purposely refrained from attempting to answer questions about "who" might have designed life. In *The Return of the God Hypothesis,* however, it has been by reported by his publisher that he will bring his ideas full circle and provide a reasoned and evidence-based answer to the ultimate mystery of the universe, drawn from recent scientific discoveries.

Also according advance publicity, Meyer will cite the following to make his case:

- Evidence from cosmology showing that the material universe had a beginning.

- Evidence from physics showing that, from the beginning, the universe was been "finely tuned" to allow for the possibility of life.
- Evidence from biology showing that since the universe came into being, large amounts of genetic information present in DNA must have arisen to make life possible.

Again, according to advance publicity, using evidence from these three fields of science, Meyer will reveal how the data support not just the existence of an intelligent designer of some kind—but the existence of a theistic creator. That is not something I attempt to do in this book, and so it will be interesting to read what Meyer has to say. Nevertheless, his earlier books, and the information that has already come out in publicity and videos about his upcoming book, make a convincing case for mind and intelligence being the ground of being of physical reality and life.

The Miller-Urey Experiment and the Discovery of the Double Helix

In the mid twentieth century, just about all scientists believed that life came about by accident, as had everything else—that there had been a primordial soup containing the right chemicals that had been struck by lightening, and somehow life had been formed as a result. This hypothesis was supported by what is known as the Miller–Urey experiment conducted in 1953, which simulated the conditions then thought to be present on the early Earth. Life was not created by that experiment, but some amino acids necessary for life did come about.

Then along came the discovery that same year (1953) by James Watson and Francis Crick of the double helix, the twisted-ladder structure of deoxyribonucleic acid (DNA). This marked a milestone in the history of science, and it gave rise to modern molecular biology, which is largely concerned with understanding how genes control the chemical processes within cells. In short order, this discovery yielded groundbreaking insights into the genetic code and protein synthesis. During the 1970s and 1980s, it helped to produce new and powerful scientific techniques, specifically recombinant DNA research, genetic engineering, rapid gene sequencing, and monoclonal antibodies, techniques on which today's multi-billion dollar biotechnology industry is founded.

An Impasse Is Reached

According to Stephen Meyer, because of this new knowledge, by the mid 1980s scientists trying to determine the origin of life had reached an impasse and there has been no progress since. There were many problems that caused this but the most fundamental one was the discovery of the information-bearing properties of DNA and biomacromolecules. In 1957 Francis Crick realized that the chemical subunits along the interior of the double helix were functioning just like alphabetic characters in a written language, or the digital characters such as the zeros and ones in a computer code, and that they are what direct the construction of proteins and protein machines that all cells need to stay alive. In other words, digital information directs the construction of the crucial components of living cells. Therefore, to explain the origin of life, one would have to explain how this complicated processing system came about. So the big question became, how can chemistry produce code?

As a result, back in the mid 1980s, a number of scientists began to see that there had to be some sort of guiding intelligence responsible for the origin of life. It became apparent to them that information in the form of code—not unlike computer code—is the key. That's when Stephen Meyer, then in his mid twenties, became interested it the origin-of-life question. He believed there must be a scientific way to answer it, and so he went to Cambridge University in England to study, and while doing so, to try to find the answer. There, he earned a Ph.D. in The Philosophy of Science.

Although he did not find the answer at Cambridge, Meyer did become convinced there was a way to make the case for a design hypothesis. He realized, however, there were still a lot of questions to be answered. One had to do with what is meant by "information," which has multiple definitions. The mathematical definition, a measure of the improbability of a sequence of symbols, is called "Shannon Information," named for Claude Elwood Shannon (1916 – 2001) who was an American mathematician, electrical engineer, and cryptographer known as "the father of information theory." Unfortunately, Shannon Information doesn't capture the notion of meaning or function, and therefore, an improbable sequence could be the result of a random process. DNA has much more than an improbable sequence, it has specificity of arrangement in order to produce a function. It is the same sort of information used in computer code, or in a book composed of letters of the alphabet that produce words and meaning such as those you are reading right now. Meyer began searching for an analytical tool to capture and identify that which indicates intelligent design as opposed to undirected natural processes. Its existence, of course, would contradict the tenet of Materialist Science that nothing in the universe exists except material substance. You see, unless it can be touched, seen under a microscope, or measured in some say, whatever someone may assume to exist does not and cannot exist, according to Scientific Materialism. Meyer knew that his hypothesis conflicted with what scientists thought about, for example, the Miller-Urey experiment mentioned above, that produced amino acids, i.e., building blocks of proteins. However, to get a protein that folds up and does an actual job, the amino acids would have to be sequenced properly, which would require a big information input, and so the unanswered question about the origin of life was, where does that information come from? Scientists back in 1953 thought they knew how life evolved, but the more those scientists learned, the more those that were paying attention realized that they did not know. In other words, the more science revealed, the more it progressed, the more the concept of a creator—or a mind—as the cause of life became plausible. Materialist Scientists long have labeled the theological perspective that gaps in scientific knowledge are evidence or proof of God's existence the "God of the gaps" argument, but perhaps to their chagrin, what was being revealed drip-by-drip was that the opposite of the "God of the gaps" argument was true. The more that scientific knowledge advanced, the more that science knowledge seemed to add credence to the existence of Infinite Mind.

Meyer Uses Darwin's Method

Stephen Meyer applied the method of reasoning he learned from Darwin in *Darwin's Doubt* and in *Intelligent Design* when making the case for intelligent design in biology. Darwin's rule was that to explain an event that took place in the past, one must identify cause and effect processes and find a *vera causa,* or true cause—in other words, a cause that is known to produce in the effect in question. What, Meyer asked, is the cause that we know produces digital information? We know of only one: a mind, i.e., a programmer. DNA is like a computer program only much more advanced that any humans have yet devised. Just how advanced and complicated is it? According to an article on the website of *BBC Science Focus Magazine,* the UK's leading science and technology monthly: "The

DNA in your cells is packaged into 46 chromosomes in the nucleus. As well as being a naturally helical molecule, DNA is super-coiled using enzymes so that it takes up less space. If you stretched the DNA in one cell all the way out, it would be about two meters long and all the DNA in all your cells put together would be about twice the diameter of the Solar System."

How incredible is that! Think of the enormous amount of information packed into DNA. Here's a link to the article just referenced:

https://www.sciencefocus.com/the-human-body/how-long-is-your-dna/

Information Has to Come from an Intelligent Source

Here's an important point: We know from experience that whenever we see information, and we trace it back to its source, whether it's computer code, a paragraph in a book, or a computer program to simulate evolution, for example, there is always an intelligent input that accounts for that information. The inference, of course, is that intelligence is behind the origin of life. A Scientific Materialist would not agree and would say that given infinite time, anything can occur. This is the idea that a room full of monkeys with typewriters will produce *War and Peace* or the complete works of Shakespeare with no typos, given enough time. What argues against this is the Big Bang theory, which is believed to have occurred 13.8 billion years ago, and the realization that the earth is only about 4.5 billion years old. As mentioned earlier, some scientists argue against the theory, but it seems to me that whether the universe has always existed and contracts and expands, the result would be the same—it got off to a (perhaps new, after an infinite number of previous) start(s) 13.8 billion years ago. This is indicated by a broad range of phenomena, including the abundance of light elements, the cosmic microwave background (CMB), large-scale structure and Hubble's law, i.e., the farther away galaxies are, the faster they are moving away from Earth.

In his upcoming book, *The Return of the God Hypothesis,* Stephen Meyer will argue that the odds are much too long for the laws of the universe that enable life to have happened by accident, and that given the age of the earth, there was not enough time for life to have occurred by chance. In other words, there is a limit on what he refers to as the "probabilistic resources," i.e., the number of tries it would take for either of the two to have happen by chance. He maintains that time is not a friend of the hypothesis that life began by a random process.

The Law of Entropy Argues Against Chance-Based Progress

And suppose, for example, a chance process does result in something that's moving in the right direction for the creation of life. Entropy, i.e., the natural direction of spontaneous change toward disorder, will work against making further progress, meaning that all the other processes will likely unwind that progress before additional progress can be made.

In summary, Meyer argues that the chance-based argument is faulty for two reasons: 1) time works against the chemical synthesis of life, and 2) there is a limit to the amount of time for it to have happened. In both of Meyer's previous books, he looks carefully at the mathematics of the chance-hypothesis for life. He will likely do the same for the laws of the universe that make life possible. For example, something that runs contrary to logic is that when water (a liquid) becomes a solid (ice), it floats. It does so because its crystalline structure makes ice lighter than water. If instead water condensed and became heavier when it became solid—like practically every other substance when it changes from a liquid to a solid except water—it would be impossible for aquatic life to survive in cold climates. If ice were heavier than water, it would sink to the bottom of oceans, lakes, and ponds. In freezing temperature conditions, it would pile up and stack up. As a result, those bodies of water would eventually become solid blocks of ice where life could not survive.

Probabilistic Resources

Using the example of a four-dial bike lock, how likely is it a thief will be able to break the four-digit code? The odds will be against it happening unless he has enough time to sample more than half of the possible combinations. Therefore, when assessing the plausibility of a random search for an informational sequence, it's necessary to assess how many opportunities there are to do so, versus the complexity of the sequence. Concerning life, it turns out that when you do the math, both in the pre-biotic case (the conditions that make life possible) and in the biological case (once life exists, that mutation and selection result in the evolution of life), the complexity of the sequences is so great—making the number of combinations that would have to be searched so large—that there are not enough of what Meyer calls the "probabilistic resources" that would be needed to build even standard length functional proteins, much less life itself. To say this another way, there are so many combinations that would have to be searched that there is not enough time for it to have happened.

As indicated by the statement above from *BBC Science Focus Magazine,* what is found in DNA is so complex, it would not be appropriate to compare it to the complete works of Shakespeare or to *War and Peace.* It   is more like all the books in the Library of Congress and then some. Meyer says that the calculations he has made show conclusively that it is overwhelmingly more likely that a random search for a new gene sequence capable of building a new functional protein would fail than it would succeed, given all the possible opportunities there were for such a search to have occurred between the time the earth formed and the first life appeared. And that's just for a functional protein, not a full-blown, living cell.

Who Is the Designer?

Life appears to be designed, and the universe also looks designed. The more we know, the more this appears to be the case. Of course, Intelligent Design doesn't posit who the intelligent designer might be. The theory of Intelligent Design posits that there are certain features of living systems that are best explained by a designing intelligence, as opposed to a purely undirected, material process. Up until his forthcoming book, Meyer's writings have been about the design we seen in life. He has not attempted to identify who was the designer. He says that's because there are two possibilities. One, of course, is God. The other is that because life arose on earth well after the Big Bang, it's possible there could be some immanent (existing or operating within, i.e., inherent) form of intelligence that might be responsible for the design that we see in cells and organisms. Richard Dawkins, the English evolutionary biologist, author, and outspoken atheist, proposed this hypothesis when he was being interviewed by Ben Stein—the conservative American writer lawyer, actor, and commentator— in the film *Expelled.* Dawkins suggested it might be a space alien or some agent within the cosmos that seeded life on planet earth. Apparently, Francis Crick, who as mentioned above determined with James Watson the double helix structure of DNA, wrote this up, I believe, in his book, *Astonishing Hy-*

pothesis: The Scientific Search for the Soul. I have to say here that I have not read that book, so I could be wrong, but in an interview, Meyer attributed the immanent designer hypothesis to both Crick and Dawkins. The problem with it, of course, is that even if it is true, the obvious question that would come next is, "Where did the information come from to form the 'immanent form of intelligence' that later seeded life on earth?" In other words, "Who created the space alien?"

Meyer has said publically that in his new book he explores what physics and cosmology can tell us about the identity of the designer. Moreover, the laws of physics, i.e., the design of the universe itself, appear to be fine-tuned not only to support life, but to allow the universe to exist—that without them, it couldn't and wouldn't. Meyer says we have what he calls a "Goldie Locks" universe where everything is just right to support life, and not only that, the laws are necessary for the universe itself to continue. He says this has been so from the very beginning—that at the moment of the big bang, the universe appears to have been pre-designed to become what it has become. The bottom line is that unless we assume some form of super intelligence as the first cause, we not only cannot explain the creation of life, we cannot explain the creation of the universe. But then again, perhaps Materialists are correct who believe that the universe has always existed and that it contracts and expands, just as we breathe in and out. That, apparently, is what Hindus believe. If true, it wouldn't negate the idea that the universe itself is an infinitely conscious and intelligent living organism—an Infinite Mind that has spawned all life and each one of us. As previously mentioned, it will be interesting to read Meyer's argument for a theistic creator.

The Two Types of Fine Tuning

Two types of fine-tuning apparently occurred. One has to do with the strength of the physical laws, i.e., the "universal constants." For example, if gravity were a little stronger or weaker, if electromagnetism were a little stronger or a little weaker, if the ratio between these forces were not what they are, life would not be possible. These forces are exquisitely fine-tuned, something like one in ten to the fortieth power, or more. An engineer will tell you, that's an incredibly tiny, tiny tolerance.

The other type of fine-tuning has to do with the arrangement of matter at the beginning, which is a separate issue from the universal constants. To get the stable and organized things we call galaxies that can host planetary systems, the Oxford mathematical physicist, Sir Roger Penrose, has calculated that the original configuration of matter would have to have been exquisitely finely-tuned. According to Meyer, the number that Sir Penrose came up with in his calculations was one in one part to the ten to the ten to the one-twenty-three. I imagine Meyer will show the math behind all that in his upcoming book, but suffice it to say as he said in a video on YouTube, "You can fill a universe with the zeros after that number."

During the past forty to sixty-plus years or so, the science that has been revealed behind the creation of the universe and the creation of life makes it more and more difficult, shall I say "painfully" difficult, to believe that it happened randomly. The only common-sense interpretation of the data that I can see, or that Meyer can see, is that all this fine-tuning requires a fine-tuner. I would imagine Stephen Meyer's new book will contain even more examples. According to Meyer, the most popular alternative proposed today by Scientific Materialists is that many, many, many universes exist, and this just happens to be the one that is hospitable to life. Of course, anyone who knows about odds understands that each time a coin is flipped, the odds of it coming up heads or tails is the same, 50-50, no matter how many times in a row previous flips have turned up heads or tails. That means the odds are the same, i.e., a few zillion or more to one, for a this universe to have been fine-tuned for life no matter how many other universes there may happen to be. Moreover, in Meyer's new book, he will argue that for the infinite-number-of-universes, i.e., multi-verse, hypothesis to be plausible, there needs to be some mechanism for generating all those universes. Besides that, if all those universes were just out there, and they have no connection to ours, they wouldn't affect the probabilities in our universe, just as the flip of a coin is not affected by the results of a flip that came before or after it. In order to portray our universe as the result of a giant lottery, there has to be some kind of common cause, some universe-generating mechanism that's responsible for all the different universes—sort of like a cook throwing spaghetti against a wall to see what sticks. Meyer will show in his new book that the universe generating mechanism—whom ever is throwing the spaghetti—would have to be fine-tuned to produce new universes, and so all it does is push back Intelligent Design one step. He will also show that since the evidence indicates Intelligent Design existed from the initial moment of the Big Bang, the implication is that this intelligence existed before the Big Bang and was likely the cause of it. God or Infinite Mind? I'll let you decide the answer to that question.

How I Became Interested in this Topic

Up through college, I believed in Scientific Materialism and was a self-described agnostic, but in my mid twenties, something happened—what you might call a paranormal or mystical experience—that raised serious doubts in my mind about the materialistic worldview I'd been brought up to hold. I became fascinated with metaphysics and joined the Rosicrucian Order, an organization that studies metaphysical law, and rose from Novice to Adept. In addition, I read everything on the subject I could get my hands on. One such book in particular changed how I viewed the world. Published in 1975, it refuted the idea that intelligence, consciousness, and awareness, came about as a result of evolution.

As previously discussed, if matter is all that is as many scientists believe today, consciousness could not have existed until evolution produced a brain. The book was entitled *Intelligence Came First,* and it was compiled and edited by Ernest Lester Smith, a

Fellow of the Royal Society—the prestigious scientific academy of the United Kingdom, dedicated to promoting excellence in science. Smith's book caused quite a bit of controversy when it came out. The premise is that, throughout the eons of evolution, needs have preceded the organs through which they are fulfilled—eyes, ears, taste buds, hearts, kidneys, and so forth. Since each new organ developed in response to a need, why would the brain be an exception? The book put forth a compelling argument that intelligence came first, quite able to function in its own realm. This book has long been forgotten, perhaps by everyone except me, because Materialists shouted it down with a vengeance, but think of the intelligence that would be required to design any one of those organs. Could all of them have come about by chance, i.e., random mutations followed by natural selection? Who that has really thought about the complexity of an eye, a liver, or a kidney could possibly think it could have happened by accident? And yet it appears that back in 1975, the scientific community did just that. Will anyone who actually reads this book and considers the evidence objectively continue to believe that? According to Steven Meyer, the likelihood that chance could produce a functioning protein is as close to impossible as anything can get, and yet there are intelligence people whose business is science that believe brains, eyes, ears, hearts, kidneys, livers, and so forth came about by chance? I find that incredibly difficult to fathom.

Many years after reading that book, in 2007, I took the opportunity to become the talk show host and producer of an Internet radio show called *The Truth about Life.* For two years I read, and over that period of time interviewed, more than a hundred authors engaged in quests for truth. Among them were medical doctors, parapsychologists, meta-physicians, quantum physicists, near death survivors, theologians, psychiatrists, psychologists, and all manner of researchers into the true nature of reality. I don't recall any of these cutting-edge individuals who actually held to a Materialist point of view, except one guest who could not produce any facts to back up his claims. All he could come up with was something to the effect that a particular claim "cannot be so because it goes against what science tells us." I found that about as convincing and a statement by an evangelical Christian that, "It can't be so because the Bible says otherwise."

Hosting that show and talking to all those people on the cutting edge has led me to believe we are in the process of shifting from the old Materialist paradigm to a new one that accepts that consciousness in the form of Infinite Mind is the ground of being and that we all come from and are still connected to the Infinite Mind.

Chapter Four
The Science of Reincarnation

Jim B. Tucker, M.D., the Bonner-Lowry Professor of Psychiatry and Neurobehavioral Sciences at the University of Virginia School of Medicine, told me in one of two interviews I had with him on my radio show that for individual cases that he and his colleagues have studied, it is possible to come up with arguments that cast doubt on reincarnation as the cause, but that reincarnation is the only viable explanation to explain all the UVA cases when viewed as a whole. It seems to me the continuation of consciousness of specific individuals has been clearly demonstrated by Jim Tucker's and his predecessor, Ian Stevenson's (1918-2007) research. Some have suggested that the existence of a "psychic reservoir," or that ESP or superpsi, might provide an explanation. But these possibilities can in no way explain what often accompanies children's memories of past lives—birthmarks that mimic the wounds that ended the past life, the strong emotion many of the children feel about their past life and the loved ones they left behind. It cannot explain food preferences, sexual orientation, phobias, and cravings on the part of some for alcohol or tobacco. If you want to know more about this, I recommend Dr. Tucker's book, *Life Before Life* (St. Martin's Press, 2005) and others he has written since.

The Mind Can Exist Separated from the Body

As was covered in some detail in Chapter One, research by UVa's Division of Perceptual Studies clearly indicates that our minds—what Dr. Tucker's predecessor, Dr. Ian Stevenson (1918-2007), called the "reincarnating personality"—must be able to exist independently of the brain and body in some sort of mental space or discarnate realm. The database at the University of Virginia indicates, for example, that about one in five children reporting past life memories also report memories of the time spent between the past life and this one. Stevenson theorized there might be an intermediate vehicle, made of "nonmaterial mind stuff" that imprints the embryo or fetus with memories of injuries or other markings of the previous body, together with likes, dislikes, and other attitudes. Rupert Sheldrake, a British biochemist, graduate of Cambridge University and former Royal Society research fellow, has set forth a hypothesis that may explain this. According to Sheldrake, the growth, development and the programmed behavior of organisms are governed by fields which exist much like fields of gravity or electromagnetism, and these fields change and evolve as a species changes and evolves. Each plant, animal and human has its own field which is part of a larger field of its species just as a radio show has its own particular frequency but is nonetheless part of the full band of radio frequencies on the AM or FM radio dial.

Sheldrake is not the only one to have come up with such a theory. A man named Harold S. Burr, Ph.D., (1889-1973) did also. Dr. Burr was E. K. Hunt Professor Emeritus, Anatomy, at Yale University School of Medicine and a member of the faculty of medicine for more than forty years. From 1916 to the late 1950's, he published, either alone or with others, more than ninety-three scientific papers. Dr. Burr maintained that all living things—from men to mice, from trees to seeds—are molded and controlled by electro-dynamic fields, which he was able to measure and map with standard voltmeters. He maintained that these "fields of life," or L-fields as he called them, are the basic blueprints of all life.

Morphogenetic Fields Work Together with Our Genes

Sheldrake's theory is essentially the same as Dr. Burr's. According to Sheldrake, genes and morphogenetic fields work together to create our bodies. Genes account for such things as hair and eye color, and other inherited features. Morphogenetic fields guide the cells of a growing fetus to become a kidney or a foot or a brain while an animal or human embryo is forming in the womb.

Traditional biology assumes genes are programmed with the purpose of each new cell and direct it to form whatever body part it is assigned to, but this has never been demonstrated. Sheldrake says genes dictate the primary structure of proteins, not the individual parts of the body. According to currently accepted theory, given the right genes and hence the right proteins, and the right systems by which protein synthesis is controlled, an organism is supposed to assemble itself. But how does this actually work? As Rupert Sheldrake once wrote, "This is rather like delivering the right materials to a building site at the right times and expecting a house to grow spontaneously."

Physiologists do their best to explain the functioning of plants and animals in mechanistic terms, but explanations of some phenomena are sketchy at best. Sheldrake believes the following can be explained by the existence of morphogenetic fields: Formation of the structure of organisms, instinctive behavior, learning, and memory.

Sheldrake's theory also clears up certain mysteries that currently remain with respect to the theory of evolution. According to the fossil record, a species can remain virtually unchanged for many millennia and then alter dramatically during an epoch when environmental conditions shift. This happens so quickly that scientists often are unable to find evidence of the transition. An eminent authority on evolution, Stephen Jay Gould (1941 – 2002), once wrote, "The extreme rarity of transitional forms in the fossil record persists as the trade secret of paleontology. The evolutionary trees that adorn our textbooks have data only at the tips and nodes of their branches; the rest is inference, however reasonable, not the evidence of fossils."

One Way Evolution Might Possibly Work

It seems a reasonable possibility that we humans and everything else in the universe evolved out of an organizing intelligence that in some of my books I have called "Spirit," and in this one,

"Infinite Mind." Perhaps life on earth began this way: In the beginning, Infinite Mind created an almost infinite number of variations of living things. Let's say they were one-celled animals in the sea. Those that were most suited to the environment survived. They reproduced by the millions, each offspring slightly different from its siblings. Imbued with intelligence, more complicated forms began to appear. Again, those best suited to the environment survived and reproduced. And so on and so on.

As evolution progressed, living organisms became more and more aware or conscious, and intelligent. This intelligence impressed itself upon the organizing intelligence of the Infinite Mind, and the organizing intelligence of the Infinite Mind went to work to create ever more sophisticated and evolved adaptations in a sort of push-me, pull-me way. If so, the result of this process can be seen in ever-increasing levels of intelligence displayed by ever more evolved life forms because, as can be seen, as intelligence evolves it becomes increasingly self-aware.

Let's say the time has come when animals that are at least somewhat self-aware are walking around on land. One such animal eats leaves and lives in an environment that's changing from forest to savanna. For millennia, plenty of leaves were available to eat in the forest, but dryer conditions are developing, punctuated by rainy spells, and much of the low-lying vegetation has died off and been replaced by grass. Some trees are able to survive the dry spells, and more and more, the leaves that remain in this changing environment are found on trees that tend to be fairly high off the ground.

Perhaps this animal walks each day by trees the leaves of which are too high for him to reach. He thinks to himself, "Doggone it, if only I had a longer neck I could feast on those leaves up there." In some way or other this remedy to the predicament forms in the animal's subconscious mind, which is an aspect of the species' morphogenetic field. Newborns of this species begin being born with longer necks as a result, and the animal we know as a giraffe develops in what would be considered a short period of time, geologically speaking. Natural selection also favors those with longer necks and works together with the morphogenetic field in a push-me, pull-me effect.

I'm not saying this is actually how evolution works. It's only a theory. Hopefully, someone smarter than I am will come up with one that's better.

Rapid Evolution Does Happen

The fact of the matter, however, is that rapid evolution does happen. A fairly recent example may be that of the Tasmanian devil, reported on in a July 15, 2008, Associated Press article. Researchers at the University of Tasmania in Australia wrote up the case. Faced with an epidemic of cancer that cuts their lives short, Tasmanian devils have begun breeding at much younger ages than one would expect.

"We could be seeing evolution occurring before our eyes. Watch this space!" zoologist Menna Jones of the university was quoted as having said.

Tasmanian devils live on the island of Tasmania, south of Australia. They weigh 20 to 30 pounds and were named devils by early European settlers because the furry black marsupials produce a fierce screech and can be bad-tempered.

Since 1996 a contagious form of cancer called "devil facial tumor disease" has been infecting these animals and is invariably fatal, causing death between the ages of two and three.

In the past devils would live five to six years, breeding at ages two, three and four, but with the new disease, even females who breed at two may not live long enough to rear their first litter.

Jones, who has been studying the animals' life cycles since before the disease outbreak, noted that there has been a 16-fold increase in breeding at age one. She reported her findings in the July 14, 2008 edition of *Proceedings of the National Academy of Sciences.*

The disease could cause the devils to become extinct in 25 years or so, she said, but this change to younger breeding may slow population decline and reduce the chance of them disappearing.

The Panda's Thumb

If you've ever been to the National Zoo in Washington, you've probably watched the giant pandas eating bamboo leaves. They take stalk after stalk and slide them between thumb and forefinger, stripping them, then popping this mouthwatering high-fiber food in their mouths. You may have wondered how these big guys got thumbs since primates are the ones with opposing digits. Pandas belong to the family *Procyonidae* (raccoons, kinkajous, et cetera) of the order *Carnivora,* one of the hallmarks of which is that all five digits on the front paw point forward and have claws for ripping flesh.

On close inspection you'll find that the panda's thumb is not a thumb at all but a "complex structure formed by marked enlargement of a (wrist) bone and an extensive rearrangement of musculature." Not having the thumb needed to make bamboo eating easy, the panda took what he had to work with and evolved one of a makeshift variety, according to biologist Stephen Jay Gould, who also wrote, "The panda's thumb provides an elegant zoological counterpart to Darwin's orchids. An engineer's best solution is debarred by history. The panda's true thumb is committed to another role, too specialized for a different function to become an opposable, manipulating digit. So the panda must use parts on hand and settle for an enlarged wrist bone and a somewhat clumsy, but quite workable, solution." Gould added, "Odd arrangements and funny solutions are the proof of evolution—paths that a sensible God would never tread but that a natural process, constrained by history, follows perforce."

Where Memories Reside

Assuming they exist, morphogenetic fields would also explain a phenomenon of memory that has neuroscientists puzzled: where it is located in the brain. One way research on this subject has been conducted is to train an animal to do something and then to cut out parts of its brain in an effort to find where the memory was stored. As Sheldrake has written, "But even after large chunks of their brains have been removed—in some experiments over 60 percent—the hapless animals can often remember what they were trained to do before the operation."

Several theories have been put forth to explain this including backup systems and holograms, but the obvious one in light of Sheldrake's hypothesis, as well as the research done at UVA, is that the memory is not in the brain at all. As was covered in the previous chapter, past-life research conducted by the University of Virginia reveals that subjects sometimes report memories from between lives when they have no physical body or brain. I asked Dr. Tucker about this. His answer was that brains are needed to recall memories, but that it appears brains are not where memories are stored.

The bottom line is this: scientists have been looking in the wrong place. To quote Sheldrake again, "A search inside your TV set for traces of the programs you watched last week would be doomed to failure for the same reason: The set tunes in to TV transmissions but does not store them." In other words, the brain is a physical link to the memory located either in your morphogenetic field, your Soul or Higher Self, or perhaps in your own little cubby in the psychic reservoir. Take your pick.

Instincts May Be Memories Housed in Morphogenetic Fields

It seems logical to me the morphogenetic fields of individual humans blend into the overall morphogenetic field of humankind. Each one affects the whole in terms of where the species stands in evolution. The same is true of species of animals. This has obvious implications in the explanation of instinctive behavior. The collective field of a species that is hunted—deer, for example—learns over time to be afraid of man. An individual deer does not have to learn this after birth. He is born with it, and we label it "instinctive behavior." It's part of the collective memory of deer that is contained in the morphogenetic field of the species.

Adherents to the "survival of the fittest" theory will argue that of the many deer that are born each spring, those that possess a natural inclination to skittishness are more likely to reach the age of reproduction, and this is what has caused the trait to develop into an instinct over time. This makes sense as well, so it's hard to argue. My guess is—since most things have more than one cause—that both theories are correct and in fact work together as written above in the hypothetical case of the evolution of giraffes.

Other Evidence of Morphogenetic Fields

The fact of a collective morphogenetic field would help explain the behavior of societal insects, fishes and birds. For example, we've all seen swarms of gnats, schools of fish, or flocks of birds behaving as though they were a single organism as they glide through the air or water, turning and diving as though they form one unified whole. Spend some time at an aquarium watching a school of fish. Something is sure to cause a minor explosion in their midst, producing momentary chaos as individuals scatter a short distance from their original positions. But within seconds, they will regroup and become a single moving organism once more.

The behavior of some species is truly amazing, or would be without Sheldrake's and Dr. Burr's theory. Key West silver-sided fish, for example, will organize themselves around a barracuda in a shape that seems dictated by risk. The distance between the school and the barracuda is widest at the predator's mouth and narrowest at the tail, where the threat of being eaten is the least.

In the world of insects, African termites, which are blind, rebuild tunnels and arches from both sides of a breach and meet up perfectly in the middle, and they can do this even when the two sides are separated by a large steel plate that is several feet wider and higher than the termitary, placed so that it divides the mound.

Acquired Characteristics Can Be Passed Along

Be all this as it may, what may be mind-blowing about Sheldrake's hypothesis to those accustomed to thinking of heredity as working solely by the passing of genes through egg and sperm is this: acquired characteristics can be passed from one generation to the next. Dr. Stevenson found that birthmarks and other physiological manifestations often relate to experiences of the remembered past life, particularly when violent death was involved. He wrote two books on this subject, reporting on a total of 225 such cases: *Where Reincarnation and Biology Intersect* (Praeger, 1997), and *Reincarnation and Biology: A Contribution to the Etiology of Birthmarks and Birth Defects* (2 Vols., Praeger, 1997).

In my interview with Dr. Tucker, he pointed out that in some situations mental images are known to produce specific effects on the body. For example, some religiously devout individuals develop wounds, called stigmata, which match the crucifixion wounds of Jesus. More than 350 such cases have been reported.

Someone under hypnosis might be told a pencil is a lit cigarette. When the pencil touches that person's arm, a cigarette burn will appear on the arm.

In another case, Dr. Tucker said a man who remembered a traumatic event when he was tied up developed rope marks on his arms. It's amazing what belief and the power of the mind can create.

Let's take a look at the awesome power of belief.

The Power of Belief

The effectiveness of placebos, for example, has been demonstrated time and again in double blind scientific tests. The placebo effect—the phenomenon of patients feeling better after taking dud pills—is seen throughout the field of medicine. One report says that after thousands of studies, hundreds of millions of prescriptions and tens of billions of dollars in sales, sugar pills are as effective at treating depression as antidepressants such as

Prozac, Paxil and Zoloft. What's more, placebos cause profound changes in the same areas of the brain affected by these medicines, according to this research. Thoughts and beliefs can and do produce physical changes—in this case in our bodies.

The same research reports that placebos often outperform the medicines they're up against. For example, in a trial conducted in April, 2002, comparing the herbal remedy St. John's wort to Zoloft, St. John's wort fully cured 24 percent of the depressed people who received it. Zoloft cured 25 percent. But the placebo fully cured 32 percent.

Taking what one believes to be real medicine sets up the expectation of results, and what a person expects to happen usually does happen. It's been confirmed, for example, that in cultures where belief exists in voodoo or magic, people will actually die after being cursed by a shaman. It appears such a curse has no power on an outsider who doesn't believe. The expectation causes the result. If you've read my novel, *The Mt. Pelée Redemption,* you know I used this phenomenon as a factor in the plot. In my book, *A Witch in the Family,* I cited this phenomenon as possibly being a factor behind the Salem witch hysteria of 1692. I believe many of the afflicted really did believe witches were tormenting them. Some developed lesions on their skin that looked like teeth marks where they thought witches had bitten them. Others had fits and coughed up blood.

No Wonder Athletes Get Better and Better

The implications of Sheldrake's hypothesis are widespread. To give an inkling of those falling outside the parameters of this book, consider this: during the past century athletes achieved ever-higher levels of excellence in everything from Olympic track and field to tennis. Improvements in diet, equipment, training techniques and coaching have certainly played a big role, but we must now also consider whether memories located in morphogenetic fields may also be a factor. According to the theory, what has been learned by the pioneers in a sport would become embedded in the morphogenetic field of humanity, and this should make learning the sport, as well as body and muscle coordination, easier for future participants. This might also account for child prodigies and virtuosos. Could it be, for example, that Tiger Woods is the incarnation of a twentieth century golfing great?

Chapter Five
Evidence Awareness Is Nonlocal

Back in the early 1930s a university with a new name and big ambitions hired a couple of men who wanted to unravel the mysteries of the paranormal. That university was Duke, located in Durham, North Carolina, now one of the most prestigious in the United States. The men were William Mc-Dougall and Joseph Banks Rhine, most often referred to as J. B. Rhine. The organization they created was called the Duke Parapsychology Laboratory for many years. Today it is called The Rhine Research Center, and although it is no longer connected with the University, it is located adjacent to the Duke campus.

What motivated these men? They wanted to prove or disprove the fact or fiction of life after death. On my radio show that aired the week of April 6, 2009, I interviewed journalist Stacy Horn who wrote a book chronicling the history of this organization from 1930 to 1960, including experiments that were conducted and the interaction of the many people over the years. This included such well-known celebrities Upton Sinclair and scientists such as Albert Einstein. The name of her book is *UNBELIEVABLE: Investigations into Ghosts, Poltergeists, Telepathy, and Other Unseen Phenomena, from the Duke Parapsychology Laboratory* (HarperCollins, ECCO Imprint, 2009). Stacy went into this project a skeptic about paranormal phenomena, but was no longer a skeptic when she came out of it.

Previously known as Trinity College, a grant by tobacco millionaire James B. Duke in 1924 prompted the name change. Perhaps, the newly reconstituted school was looking for ways to make its mark when it lured William McDougall from Harvard University to set up a department of psychology.

He was soon contacted by a man named John Thomas who had 800 pages of transcripts generated by mediums he had been working with. Thomas' wife had died unexpectedly during an operation, and Thomas had been devastated. He began working with mediums in order to communicate with her.

Thomas got exciting results, but he wasn't sure he could believe them. Looking for verification of their authenticity, he traveled around the United States talking with mediums. He went to Europe, eventually, reasoning that mediums there would have no way of knowing anything about him or his wife. If they were able to come up with information that was accurate, it would be more convincing.

Ultimately, Thomas wrote to McDougall asking if he could send J. B. Rhine, then of Harvard University, and Rhine's wife Louisa, to Duke to study this material. McDougall agreed and Rhine came to Duke.

J. B. Rhine Takes Up Residence at Duke

Rhine studied Thomas' transcripts. He was able to verify much of the information, and to all but eliminate fraud and lucky guesses. He traveled to Upstate New York, for example, investigating cemetery head stones to check out the veracity of genealogy of Thomas' wife indicated by a medium. The genealogy proved to be accurate. Not even Thomas himself knew if this genealogy was correct, but the information did check out. Ultimately, however, Rhine concluded that even though the information was correct, it could not be said with absolute certainty that the information was coming from Thomas' deceased, and now disembodied, wife.

The problem still dogs researchers who study the purported abilities of mediums. Assuming no fraud is being perpetrated, several possibilities exist as to the source of information coming from mediums that seems to be from a deceased individual, all of which indicate that Scientific Materialism's contention that thoughts and mind are confined within the brain is not valid:

- It may actually be coming from the now disembodied individual.
- The medium may be employing ESP or telepathy to read the minds of living individuals—in this case Thomas himself, or other living relatives of his wife.

  Indeed, a whole range of psychic abilities may be put to use including remote viewing, psychometry and clairvoyance. Nowadays, the full breadth of psychic abilities that might be at work is called "superpsi."

- A third possibility is that the medium might be tapping into a reservoir of information of human history, thoughts and feelings many believe exists. Some call this the Akashic Records, which are envisioned as the memory hard drive—or perhaps "the cloud" would be a better metaphor—of the universe. The famous psychiatrist, Carl Jung, for example, wrote of a universal unconscious that holds the history and thoughts of all mankind. Today, researchers call this the "psychic reservoir." This is thought to be the source of information for perhaps most famous and well-documented psychic of the twentieth century, Edgar Cayce (1877-1945), often referred to as "The Sleeping Prophet"—Cayce became known as such because his readings were given while in a self-induced trance.

Rhine could find no way to prove superpsi, or the psychic reservoir, were not the source of information tapped into by mediums that had supposedly been in touch with Thomas' wife. So Rhine began putting his energy into the study of what became known as extra sensory perception, or ESP. He reasoned that if he could prove awareness extends beyond, and exists outside the body, a major step would be taken toward establishing the possibility of survival of consciousness after death. After all, for our consciousness to continue after death it has to be capable of existing outside the body and the brain, which we already know is true because of findings by the Division of Perceptual Studies at the University of Virginia.

Strong Evidence of ESP

Rhine's most famous experiment used what has become known as ESP cards. Developed specifically for this purpose, these had different symbols on them including a star, wavy lines, a cross, a box and a circle. Many of these experiments were conducted—mostly using Duke University students—to see if people could tell what symbols were on the cards without looking at them. It was found again and again that they could.

The controls employed in these experiments were refined over time until neither the students nor those testing them could see one another. Ultimately, research was conducted in such a way that not even the person conducting the experiment knew what symbol was on the card a student was to identify. The experiments turned up statistically significant results time after time, showing without a doubt ESP is real.

One of Rhine's subjects in the ESP experiments was particularly impressive. A divinity student, his name was Hubert Pierce. Rhine believed that everyone possessed psychic abilities, but his research indicates some people have more talent for it than others. This is of course true of other abilities. An extremely talented singer will wow the judges and go on to win American Idol, but most will fail miserably and get the boot at the first audition.

There were twenty-five cards in the ESP deck, and five different symbols. Therefore, one would expect to guess five correctly each time through, simply by chance. Hubert Pierce could consistently get more than five correct, as could a number of others. But the interesting thing is, and according to Stacy Horn this came up frequently in the research, emotions played a role. Hubert, for example, needed money. He was a poor, struggling college student. Rhine once told him if he got the next card right, he'd pay him a hundred dollars. Pierce got it right. Rhine said, "Okay, get the next one right, and you'll get another hundred dollars."

Pierce got the next one right.

This went on through the entire deck. Pierce named all 25 cards correctly.

At one point, however, Hubert said he would not be coming into the lab for tests. His girlfriend had broken up with him, and he was heartbroken.

When he finally did come back, he did not perform well.

Another example of emotions playing a roll was the time Rhine tested the psychic abilities of children at a orphanage. One little girl became quite attached to a woman researcher. The little girl performed extremely well, apparently because she was eager to please, and wanted to prolong the session.

Research that Indicates Vegetation Is Conscious

Apparently, living plants are conscious. Scientifically constructed, double blind experiments by researchers, including theoretical biophysicist of the University of Marburg in Germany, Fritz-Albert Popp, have demonstrated this. And this isn't news. About 50 years ago a fellow named Cleve Backster (1924-2013) demonstrated plants are aware by using polygraph machines. In Backster's most famous experiment, he hooked up plants in his office suite to polygraph machines, and then set up a device to randomly dump a cup of living brine shrimp into a pot of boiling water. The needles on the polygraph machines would go wild each time the shrimp hit the water and went to their deaths. The plants were picking up their distress and demise.

But what led Cleve Backster to construct and carry out this experiment may be even more of an eye-opener. Lynne McTaggart, author of *The Field: The Quest for the Secret Force of the Universe,* told the following story on my radio show early in 2008.

Backster was an expert on polygraph machines and their operation—in other words, lie detectors. One evening, when Backster was a young man, he was sitting in his office with nothing much to do. His eyes fell on an office plant and he had an idea. He decided to hook up one of his machines to the plant and see if he could get it to react. He connected the machine and poured a glass of water into the soil around the plant. Nothing happened. The polygraph registered boredom.

Backster started thinking about what he might do to get a reaction out of the plant, and he had an idea.

"I think I'll burn one of its leaves."

At that moment, the polygraph machine went wild. The plant had reacted to his thought! The more Backster thought about burning the plant, the more the needle on the polygraph machine went ballistic.

Cleve Backster conducted many experiments along these lines which are described in his book, *Primary Perception: Bio Communication with Plants, Living Foods, and Human Cells* (White Rose Millennium Press, 2003).

People who have what's called green thumbs may think it is because they send kind thoughts to their plants. It may be true that kind thoughts help make happy plants, but as we now know, thoughts are not sent and received. Thoughts just are—part of the mind we and everything and everyone share.

The Phenomenon of Remote Viewing

Something else that demonstrates awareness is nonlocal—at no particular place but everywhere at once—is the phenomenon of remote viewing. Those adept at remote viewing can direct their consciousness to be anywhere they want it to be.

Remote viewers use psychic powers to observe what's happening at a location some distance from them—in terms both of miles and in some cases, time as well.

Back in the 1970s, the U. S. government learned that the KGB was using psychics to spy on the United States. Naturally, U.S. Intelligence leaders wanted to see if this actually worked.

Did it? U.S. Army Major General Edmund R. Thompson, who was deputy Director for the Management and Operations for Defense Intelligence from 1982-84 is quoted as having said, "I never liked to get into debates with the skeptics, because if you didn't believe that remote viewing was real, you hadn't done your homework."

Remote viewing was used from the early 1970s forward through the Cold War to keep tabs on what the Soviets and Eastern Block countries were up to that couldn't be observed by spy planes, or satellites, or operatives on the ground.

In Spring 2009, I interviewed F. Holmes Atwater, the man who in 1979 set up the U.S. Army Intelligence unit responsible called Stargate. His friends know him by the name of "Skip."

Skip got into this line of work through a series of what some people might call amazing coincidences, and others would say are synchronicities— events that look like coincidences, but seem to happen for a reason.

Skip grew up in a home with parents that took such things for granted. It was the sort of thing they talked about at the dinner table. As a kid, Skip would go off on out-of-body trips almost nightly. He related one specific story to me and my listeners to illustrate this. He was seven or eight years old at the time, and it had to do with a problem he had with bedwetting.

"It was embarrassing," he said. "I had a big, brown piece of rubber on my bed so I wouldn't ruin the mattress. My parents didn't scold me, but they did make me responsible for washing my own sheets.

"I can remember distinctly waking up one night, and I was all wet. I was screaming in anger, and my mother came in and said, 'What's wrong? Did you fall out of bed?'

"I said, 'No, I remember, I got up, and I went down the hall to the bathroom, and I sat down. And the minute I started to pee, I woke up here in bed, and I'm all wet.'

"I was mad as the dickens, and my mother hugged me and said, 'Oh, that's all right, don't worry about it. You know, Skip, sometimes you're in your body and sometimes you're out of your body, and you have to remember that when you're going to the bathroom, make sure you're in your body.'

"[What she said] made perfect sense to me, and I said, 'Oh, now I understand,' and that was the end of my bedwetting."

Atwater Learns of Remote Viewing

Skip was in the Army working for Army Intelligence when he came across a book called *Mind Reach* by Russell Targ and Harold E. Puthoff of the Stanford Research Institute. The book explained remote viewing, which didn't seem at all unusual to Skip given his experiences as a child. Naturally, a person could see things at a distance, using his mental powers. It was as though a light had suddenly flicked on. He instantly realized this could be used to gather intelligence.

At the time, Skip was in counter intelligence. It was his job to defend against wiretaps, bugging devices, and other forms of intelligence gathering by the enemy. No one in his counter intelligence unit had thought about remote viewing as a way the enemy might be spying on us. So Skip went to his commanding officer, a Colonel Webb, and gave him the book. After the Colonel had read it, Skip asked him if this remote viewing was being used on our side.

The Colonel had no idea. He thought if anything was going on, the Pentagon would be the place to find out. So he had Skip transferred to the Pentagon to take a position where he'd be in charge of a counter intelligence team. Skip would have the access he needed to find out about remote viewing and what if anything was being done about it to prevent the enemy from using it.

Before Skip was able to relocate to Washington, however, he received a change of orders. He was told to report to Fort Meade in Maryland. This was a better location for a young Army officer with a wife and children, which Skip had, because Fort Meade had family housing and good schools. It would be a much more affordable and pleasant place to live than Washington, D.C.

Documents Reveal U. S. Interest in Remote Viewing

At Fort Meade, Skip was assigned to what was known as a SAVE team—Security Activity Vulnerability Estimate team. The job was to go to sensitive U.S. installations and try to penetrate them in any way possible—as the enemy might in order to gather intelligence. Then the team would make a report to the commanding officer and provide recommendations for improving security.

Skip moved into his new job and was assigned an office that had just been vacated. The file cabinet and most of the desk drawers had been cleaned out, and an office safe had been emptied, but he did come across three documents in a bottom drawer of the desk that turned out to be classified. They reported on remote viewing experiments taking place in the Soviet sphere, funded by the KGB.

Skip took the documents to his supervising officer, a Major Keenan.

The Major looked at them. "Oh, yes, I remember these," he said. "The Lt. Colonel was very interested in this subject. Do you know anything about it?"

"Why, yes, I do, Major."

The Major took a moment and sized up Skip. "Lieutenant," he said, "from now on you're in charge of it."

And that's how Skip got his wish and started on a ten year career that eventually put him in charge of a remote viewing unit of the Army.

Atwater Learns about Remote Viewing

Skip learned that basic research had been underway since 1972 to check the validity of the Eastern Block experiments. The initial question had been whether reports of success were valid. It might be the Soviets were falsifying the results to create fear. The Stanford Research Institute had been retained to replicate the KGB experiments. To the surprise of our intelligence community, the results were positive.

By the time Skip got involved, the CIA and other U.S. intelligence agencies had been using psychics for some time to gather information, including well-known psychics such as Ingo Swann (1933-2013), who later wrote several books on remote viewing. Skip's job became to set up, recruit and train remote viewers for U.S. Army Intelligence who may or not have had prior experience using psychic abilities. He developed a screening process, and for those who made the cut, a training program employing methodologies gleaned from accomplished remote viewers.

Skip's efforts met with success, but after a while he began looking for ways to enhance the results his remote viewers were achieving. This led him to The Monroe Institute (TMI) in Virginia, where he later worked as Research Director.

The Monroe Institute Proves to Be a Resource

Robert Monroe (1915–1995) had spent a career in broadcasting, culminating as a vice president of NBC Radio. After leaving NBC, Monroe became known for his research into altered states of consciousness. His 1971 book *Journeys Out of the Body* is said to have popularized the term "out-of-body experience," or OBE.

Monroe's original objective had been to develop a process by which people could learn effortlessly—while they were asleep. He developed sound patterns that would help people reach a state he called "Stage Two Sleep" and then hold them in that state. Monroe experimented on himself and exposed himself to many varieties of sound. One night in 1956, quite unexpectedly, he found himself floating over his body. He panicked and thought he must be dying. Later, he consulted medical doctors and psychiatrists about this, and eventually he understood he wasn't dying—that this experience was fairly common. As a result, he conducted more experiments to learn how to replicate what he had done, and to control it.

By the time he came to Skip's attention, Monroe had established The Monroe Institute southwest of Charlottesville, Virginia, where the public could come to share in these sound-created experiences. Skip decided to investigate, and traveled from Fort Meade to Virginia to meet Monroe.

Skip, of course, was running a secret program for the U.S. Army and could not disclose the real reason for his visit. But he did explain to Monroe that he was interested in the work being done, had read his book, and had had out-of-body experiences as a child.

Monroe invited Skip to come into his laboratory. He took him to a room that had been set up and equipped for his experiments. He had Skip lie down. Skip became nervous. He was, after all, an intelligence officer on a surreptitious mission.

"What are these sounds I've heard about—these hemi-sync® sounds?" Skip asked.

"Oh, nothing to worry about," Monroe said. "I'll just play some music at first to calm you down."

As soon as Skip was lying down on the bed with the headphones on, the door shut and the lights went out. He wondered what he'd gotten himself into.

Music came through the speakers. It turned into the sound of surf against the shore. This reminded Skip of happy childhood days spent playing at the beach.

Then droning sounds came on in the background and quite unexpectedly the bed began to rise off the floor as though it were being lifted by hydraulics the way a car in a service station is lifted for an oil change.

Skip thought, "Wow, this is a very special bed. They must have one of those lifts underneath it to push it up in the air."

As he was thinking about what must have been done to build it—the building had to have been constructed around it—he began to feel himself moving in a different direction. He seemed to be headed laterally, rather than up. That's when he realized it must not be a lift he was on. Yet the feeling was very strong, quite visceral, as though he were on an airplane circling into a landing approach. He saw or imagined that he was moving through a rock or crystal tunnel of some kind. Then he heard a voice.

"Whoa, there. What's happening, kid?" It was Robert Monroe.

"Well, I seem to be going some place," Skip said.

"Well, now, where're you going, kid?"

"I don't know," Skip answered.

Skip traveled along the tunnel, or corridor, and eventually came out of it in vast, open, white space. He said it was a little like being in a white cloud except there was no mist or fog. Everything was white, boundless, and there were no walls.

Perhaps the strangest part was that Skip watched himself arrive.

He thought, "Gosh, I've come all this way only to find I'm already here."

Skip said in our interview, "It sounds trite to say wherever you go, there you are, but that's exactly what happened to me."

He remained in the white space for a while. Then he heard Robert Monroe's voice again:

"What's happening?"

Skip was embarrassed because he'd forgotten he was in Monroe's laboratory lying on a bed.

He said, "Oh, nothing much."

"Okay . . . well, it's time for lunch."

This didn't make sense, but that didn't matter because Monroe changed the sounds coming through the headphones, and Skip felt the bed being lowered down to its original position. In a short time, the door was open and the lights were on.

Monroe was standing in the doorway.

Skip leaned over and looked under the bed.

"Oh, did you lose your wallet down there?" Monroe asked.

Skip was looking for the hydraulic lift, but there was none.

As a result of this experience, he learned there was definitely something to the sound technology Robert Monroe had developed, and the Army entered into a classified contract with Robert Monroe to do some training.

The Amazing Abilities of
Joe McMoneagle

One man Monroe trained was perhaps the most outstanding remote viewer in the Army. His named is Joe McMoneagle.

Joe had been in intelligence before joining Skip's unit. His personal story is fascinating and was related to me by a guest on my show who'd gotten to know Joe over the years through an association with The Monroe Institute.

In the early 1970s, Joe was the target of a successful assassination attempt while in the Army stationed in Germany, working as an operative in intelligence. Poison was the method. He was meeting with an intelligence contact at a restaurant, having dinner, when he felt nauseous. He excused himself and went outside to get some air. He walked around for a moment, and then saw a crowd gathered just outside the door. He went to see what the commotion was about, looked through the crowd, and could make out a body lying on the street.

People were saying, "He's dead, he's dead!"

Joe came closer and was shocked to see the body was his own.

Testing later showed he'd been subjected to a binary poison, one which becomes toxic when combined with another substance. This had allowed his assassin to slip him the poison and make his getaway before Joe sat down to dinner and consumed whatever had triggered the toxicity that killed him.

McMoneagle's consciousness, after viewing his body lying on the street, went toward the light and through the tunnel described by other near-death survivors. As is now considered typical in these cases, he arrived at a place where he was met by spiritual beings. There, he underwent some instruction and a life review.

We would know nothing of this if Joe's body had not been resuscitated. His recovery and recuperation took quite some time.

What happened that evening changed Joe in several ways. He'd had psychic experiences before his NDE, but had kept them to himself. He no longer did. He also began to have spontaneous out-of-body experiences he was unable to control.

Joe's case came to the attention of two physicists at the Stanford Research Institute, Russell Targ and Harold Puthoff. They'd already been working on a government contract to study the ramifications of the quantum mechanics theory of non-locality of consciousness. These were the same experiments described in the classified document found by Skip Atwater, and the same two men who'd authored the book he'd read.

Joe became the first remote viewer directly on the government payroll. In the course of his career in the Army as a remote viewer, Joe worked on more than 200 missions, many of which were reported at the highest levels of the U.S. military and government. Some of the information was considered so crucial, vital and unavailable from any other source, that he was awarded the Legion of Merit for his work, the second highest award the Army can give to someone in the military during peacetime.

Skylab's Fall to Earth Is Accurately Predicted

One such mission was to determine the time and the location Skylab would fall to earth. Depending on how old you are, you may recall Skylab—literally a scientific laboratory in orbit around the earth, put there for astronauts to conduct experiments in space. Launched in 1973, it weighed about 100 tons.

By 1979 its orbit was decaying and Skylab was expected to come down. The question was, "Where?"

A hundred ton metal object falling on a heavily populated area such as New York, Tokyo or London, for example, would cause a tremendous death and destruction. Super computers were enlisted to answer the question, but too many variables existed for the technology of the day. The results were unreliable.

Joe McMoneagle, Ingo Swann and a third individual, a woman whose name I have been unable to uncover, were contracted with individually to come up with an answer. None of the three knew the others were involved. All picked the same day, July 11, 1979, and almost the same time. Each was within five minutes of the other two. This was nine and a half months before Skylab came down. In addition, they all picked a location in western Australia within five miles of one another—a remote, uninhabited area.

Skylab came down there, all right, almost precisely as predicted, demonstrating awareness is not located just inside our skulls, nor is it limited in time and space—more evidence my one-mind theory is correct.

The Capture of Saddam Hussein Is Seen Six Weeks Ahead

Another startling example that awareness is nonlocal comes from a book by Stephan A. Schwartz, *OPENING TO THE INFINITE: The Art and Science of Nonlocal Awareness* (Nemoseen Media, 2007). Mr. Schwartz was on my radio show in the summer of 2008. One of the amazing stories he told was about the predictions made by a college seminar class about the capture of Saddam Hussein. On November 2, 2003, after being taught the basic skills of remote viewing, 47 of those who'd attended the seminar agreed to "Describe the location of Saddam Hussein at the time of this capture or discovery by U.S. or coalition forces." The students' data was collected and analyzed, including points of consensus concerning the physical location, as well as things that were not likely to be predictable—such as Hussein's appearance on the day of his capture. The data were photocopied and distributed to a number of people, and then turned over to a third party, Herk Stokeley, Director of Atlantic University. Stokeley placed the data in an envelope, which he sealed in front of a notary, who affixed her seal across the envelope's flap. The envelope was then placed in a vault.

Hussein was captured about six weeks later, on December 13, 2003. The remote viewing documents in the safe said he would be beneath an ordinary looking house on the outskirts of a small village near the city of Tikrit, and that the house would be part of a small compound that's bordered on one side by a dirt road and, on the other by a nearby river. Two large palm trees would mark the ends of the house. All this turned out to be true.

Remote viewing also predicted Hussein would be found crouching in a subterranean room or cave reached by a tunnel. This was true.

Remote viewing said Hussein would look like a homeless person with dirty rough clothing, long ratty hair and a substantial and equally ratty salt and pepper beard. This was true.

Remote viewing said he would have only two or three supporters with him at the time of his discovery. He had two.

Remote viewing said he'd have a gun with him. He had a pistol.

Remote viewing said he would have a quantity of money. He had $750,000 in cash.

Remote viewing said he would be defiant, but would not put up any resistance and would be tired and dispirited. This was true.

What's the take-away from all this? The one mind we all share contains all—past, present and future.

Chapter Six
Prayer Works

Here's something atheists and Scientific Materialists are going to have a really hard time with. Prayer appears to double the success rate of in vitro fertilization procedures that lead to pregnancy, according to a study published in the September, 2001 issue of the *Journal of Reproductive Medicine.* The findings reveal that a group of women who had people praying for them had a 50 percent pregnancy rate compared to a 26 percent rate in the group of women who did not have people praying for them. In the study, led by Rogerio Lobo, chairman of obstetrics and gynecology at Columbia University's College of Physicians & Surgeons, none of the women undergoing the IVF procedures knew about the prayers on their behalf. Nor did their doctors. In fact, the 199 women were in Cha General Hospital in Seoul, Korea, thousands of miles from those praying for them in the U.S., Canada and Australia. According to Dr. Lobo, "The results were so highly significant they weren't even borderline. We spent time deciding if it was even publishable because we couldn't explain it."

The fact is that after it was published, skeptics did everything they could to discredit the study, from attempted character assassination of one of the researchers involved to questioning the methodology to do with how the actual praying was carried out. But from what I can determine, none of their tactics was more than a thinly veiled attempt to hang on to that erroneous scientific tenet that brains create thought and consciousness and that it remains inside skulls.

It should also be noted that this is not the only study to indicate that prayer can have a significant effect on matters of health. Another example comes from Randolph Byrd, a cardiologist, who over a ten-month period used a computer to assign 393 patients admitted to the coronary care unit at San Francisco General Hospital either to a group that was prayed for by home prayer groups (192 patients), or to a group that was not prayed for (201). This was a double blind test. Neither the patients, doctors, or nurses knew which group a patient was in. Roman Catholic as well as Protestant groups around the country were given the patients' names, and some information about their conditions. The various groups were not told how to pray, but simply were asked to do so every day.

The patients who were remembered in prayer had remarkably different and better experiences than the others. They were three times less likely to develop pulmonary edema, a condition in which the lungs fill with fluid; they were five times less likely to require antibiotics. None required endotracheal intubation (an artificial airway inserted in the throat), which twelve in the un-prayed-for group required. Also, fewer prayed-for patients died, although the difference between groups was not large enough to be considered statistically significant.

A third study indicating that prayer and faith may have positive health effects was published in 2002 in the *International Journal for Psychiatry in Medicine*. A team from the University of California at Berkeley found that Christians and Jews who regularly attended services lived longer and were less likely to die from circulatory, digestive and respiratory diseases. Devotees of Eastern religions were not surveyed. The study examined links between religious attendance and cause-specific mortality from 1965 to 1996 in 6,545 residents of Alameda County, California. Even after adjusting for variables like health and frequency of exercise, religious devotees lived longer without succumbing to disease.

"At this point it's a puzzle why there should be this pattern," said the study's author, Doug Oman, Ph.D., a lecturer at Berkeley's School of Public Health. "It's likely a stress-buffering resource. Regular attendance at services can give people an inner peace that is unshakable. That results in less wear and tear on their bodies."

It is not surprising Dr. Lobo of Columbia University and Dr. Oman of Berkeley are puzzled by the results of their own studies. These men were schooled in Scientific Materialism. If you have believed it all your life, it isn't easy to give up the idea that consciousness is the result of electrons jumping across synapses and that thought remains inside the skull. We, of course, know why and how prayer works. It does so because all things, including people and their bodies, are products of the Infinite Mind, and as I have discussed and elaborated upon in other books I have written, the subconscious, subjective mind we all share is diligent in its effort to create whatever a conscious mind fervently believes to be true. The belief of those praying that their prayers will be answered is impressed upon the mind we all share, and the mind we all share faithfully acts upon the bodies of those being prayed for.

The Spindrift Papers

An organization exists that has as its purpose the study of what prayer techniques produce the best results. It's called Spindrift and was founded by Christian Science practitioners who have been at this since 1975. Resting next to my keyboard at this moment is a document an inch thick, printed on both sides of standard letter-size paper called, "The Spindrift Papers" (see www.spindriftresearch.org). It gives detailed information of prayer experiments conducted under rigorously controlled conditions.

The first question Spindrift researchers sought to answer is, does prayer work? The answer as we already know, is yes. In one test, rye seeds were split into groupings of equal number and placed in a shallow container on a soil-like substance called vermiculite. (For city dwellers, this is commonly used by gardeners.) A string was drawn across the middle to indicate that the seeds were divided into side A and side B. Side A was prayed for. Side B was not. A statistically greater number of rye shoots emerged from side A than from side B. Variations of this experiment were devised and conducted, but not until this one was repeated by many different Christian Science prayer practitioners with consistent results.

Next, salt was added to the water supply. Different batches of rye seeds received doses of salt ranging from one teaspoon per eight cups of water to four teaspoons per eight cups. Doses were stepped up in increments of one-half teaspoon per batch.

A total of 2.3 percent more seeds sprouted on the prayed-for side of the first batch (one teaspoon per half-gallon of water) than on the un-prayed-for side (800 "prayed-for" seeds sprouted out of 2,000, versus 778 sprouts out of 2000 in the not-prayed-for side). As the dosage of salt was increased the total number of seeds sprouting decreased, but the number of seeds which sprouted on the prayed-for sides compared to the un-prayed-for sides increased in proportion to the salt (i.e., stress). In the 1.5 teaspoon batch, the increase was 3.3 percent. In the 2.0 teaspoon batch, 13.8 percent. In the 2.5 batch, 16.5 percent. In the 3.0, 30.8 percent. Five times as many prayed-for seeds in the 3.5 batch sprouted (although the total number which sprouted was small as can be seen from the chart below). Finally, no seeds sprouted in the 4.0 teaspoon per eight cup batch.

Salt	Control/Grown		Prayed-for/Grown		% Increase
1.0	2,000	778	2,000	800	2.3
1.5	3,000	302	3,000	312	3.3
2.0	3,000	217	3,000	247	13.8
2.5	3,000	454	3,000	528	16.3
3.0	3,000	52	3,000	68	30.8
3.5	2,000	2	2,000	10	400.0
4.0	3,000	0	3,000	0	0.0

What this says is what people lying in a ditch with bombs going off around them have always known: the more dire the situation, the more helpful prayer will be. Up to a point. There comes a time when things are so bad nothing helps.

This experiment was also conducted using mung beans. The solution of salt and water ranged from 7.5 teaspoons per half-gallon of water to 30.0 teaspoons per half-gallon. The increase in the number of sprouts for the prayed-for side ranged from 3.3 percent to 54.2 percent.

Next an experiment was constructed to determine whether the amount of prayer makes a difference. This involved soybeans in four containers. One container was marked "control" and not prayed for. The other three were marked X, Y, and Z. In each run of the experiment, the X and Y containers were prayed for as a unit, and the Y and Z containers as a unit. So, Y received twice as much prayer as either X or Z. The Y container also had twice as many soy beans germinate. The results were in proportion to the amount of prayer.

Studies similar to this have been and are being carried out by a consortium of scientists put together by Lynne McTaggart, author of the books previously mentioned, including *THE FIELD: The Quest for the Secret Force of the Universe* and *The INTENTION EXPERIMENT: Using Your Thoughts to Change Your Life and the World.* When she was on my radio show in early 2008, she described some of these experiments and the terrific success she and her colleagues have had and said several of these studies were being prepared for publication.

What the prayer group studies show is reminiscent of a principle set forth by Napoleon Hill in his perennial bestseller, *Think and Grow Rich.* He wrote this granddaddy of all self-help books in the 1930s and updated it in 1960. One chapter is devoted to the principle he called "The Master Mind." Hill suggested that whatever project or purpose or goal an individual had, it could be advanced and achieved most readily by bringing together a group of people who could apply their unified brainpower to it. Hill never used the word prayer nor did he suggest people sit around and pray. But he did liken a group of minds at work on a project to a group of storage batteries connected together in a series to produce much more power than any single battery possibly could on its own. He wrote, "When a group of individual brains are coordinated and function in harmony, the increased energy created through that alliance becomes available to every individual brain in the group." He cited several examples, including the remarkable successes of Henry Ford and Andrew Carnegie, each of whom had a group of colleagues around him working and pulling together on common goals. Hill wasn't referring only to innovative thinking that leads to marketing and sales results. He was talking about much more, of creating an aura that leads to favorable events taking place, or to what might be considered by materialists as "good breaks." The Master Mind creates a force with a life of its own, a force I call "grace." This is a force that works much like unseen hands. A case Hill cited was that of Mahatma Gandhi, who led the successful non-violent revolution that freed India from British Colonial rule. Hill wrote, "He came to power through inducing over two hundred million people to coordinate, with mind and body, in a spirit of harmony, for a definite purpose."

What else have studies revealed about prayer? In the Columbia study, the people praying were from Christian denominations and were separated into three groups. One group received pictures of the women and prayed for an increase in their pregnancy rate. Another group prayed to improve the effectiveness of the first group. A third group prayed for the two other groups. According to the authors of the study, anecdotal evidence from other prayer research has found this method to be most effective. The three groups began to pray within five days of the initial hormone treatment that stimulates egg development, then continued to pray for three weeks.

Besides finding a higher pregnancy rate among the women who had a group praying for them, the researchers found that older women seemed to benefit more from prayer. For women between 30 and 39, the pregnancy rate for the prayer group was 51 percent, compared with 23 percent for the non-prayer group. This would seem to parallel the Spindrift study in that those who needed help the most, up to a point at least, saw the biggest gains from prayer versus no prayer. With Spindrift, it was salt-soaked rye and mung beans. With the Columbia study, it was older women.

Is there anything more the Spindrift researchers learned which would be helpful to know?

The quality of prayer is a factor in how effective it is, as is the quantity or amount of prayer. Like anything, practice makes perfect. More experienced practitioners got better results than less experienced practitioners. Get in the habit of praying. Do not save the practice of it only for the times bullets are flying overhead, or the airplane you're on goes into a tailspin.

The Spindrift research gives more clues on how best to pray. First, you need to know what you're praying for. Some experiments were conducted in which the prayer practitioner was kept in the dark about the nature of the seeds he was praying for. He or she did not know what kind of seeds they were or to what extent they had been stressed. Results showed a drastic reduction in the effect. The researchers concluded that the more the person praying knows about that which is being prayed for, the greater the effect of the prayers.

Another experiment measured the efficacy of "directed" versus "non directed" prayer. Directed prayer was that in which the practitioner had a specific goal, image, or outcome in mind. He attempted to steer the seeds in a particular direction. A parallel in healing might be for blood clots to dissolve or for cancer to isolate itself in a particular place in the body, i.e. where it could be cut out. In the seed germination experiments it was praying for a more rapid germination rate. Non-directed prayer used an open-ended approach in which no specific outcome was held in the imagination. The person praying did not attempt to imagine or project a specific result but rather to ask for whatever was best for the seeds in an open-ended spirit of "Thy will be done." Both approaches worked, but the non-directed approach appeared to be more effective, in some cases producing twice the results.

Using the non-directed approach is bound to conflict with the beliefs of many who hold that one must visualize a specific result and hold it in his mind. No doubt in some cases that works. The problem is that we humans often do not know what the best outcome of any given situation might be. The theory put forth by Spindrift researchers is that prayer reinforces the tendency of an out-of-balance organism to return to balance, or harmony. As we know, that's something nature naturally seeks. Empirical evidence suggests that the goals of nature are harmony and growth, and prayer supports harmony and growth. To quote the Spindrift research document, "If the power of holy prayer does, indeed, heal, then such a power will be manifest as movement of a system toward its norms since healing can be defined as movement toward the optimal or 'best' conditions of form and function." The Spindrift researchers did not try experiments in which prayer was used to try to prevent seeds from germinating. If they had, and if what they say above is true, that would not have worked.

Here's what makes sense to me: Our subconscious minds are connected to the Infinite Mind, which has access to and is immersed in all the information needed in any situation. The best result may be the opposite of what we expect, which means that it's always best to put things in the Infinite Mind's hands and simply to pray for the best possible outcome. Then consider that this best outcome has already been accomplished. Do not attempt to explain to Infinite Mind what course it should take to arrive at the best outcome, nor even what you think the best outcome would be. Let Infinite Mind decide and then find the best way to get there.

Let's sum up what we need to keep in mind about prayer:

- Belief is a key. Believe that what you pray for already exists and is in the process of manifesting.
- Practice makes perfect, or in other words, experienced prayer practitioners receive the best results.
- Quantity is a factor. More prayer brings more results.
  The more a person or group knows about the subject of their prayers, the better.
- If the best outcome is clear, visualize it, and pray for it. Consider it an accomplished fact. But do not tell Infinite Mind how to arrive at this outcome. Let it find the way.
- For positive results, the purposes of the universe need to be served by our prayers. This includes growth and development, the healing of the psyche, or in the case of physical healing, the bringing of a stressed body or physical system into harmony.

Chapter Seven
High Time for a Paradigm Shift

One reason it's hard to accept that we are all part of one mind, that consciousness is nonlocal, that brains are receivers and our body's interface with our individual consciousness, and that our consciousness may well continue after death, is that it requires scrapping the materialistic model, which many of today's scientists are highly invested in. As stated in the Author's Note, the 19th century German philosopher Arthur Schopenhauer [1788-1860] is quoted as having said, "All truth passes through three stages: First, it is ridiculed. Second, it is violently opposed. Third, it is accepted as self-evident." From my experience, I can say that ardent Materialists still appear to be in the "ridicule-to-violently-oppose" stages. But the fact is, a lot of people have already scrapped the materialist model and accepted at least certain aspects of what clearly appears to be the truth. What I've attempted to do in this book is present a new model of reality that reasonable, thinking individuals can accept based on the evidence.

It's important to understand that models are just that—approximations of how things are.

Whether we realize it or not, we each have a world-view—or a model of how things work. As I have written in other books, you might think of a world-view or model as a stack of cans that forms a pyramid you might see as a grocery store aisle end display. Each can represents an individual belief. Each belief in the display supports other beliefs. Try to change a foundational belief, and the whole thing may come tumbling down. In the past 40 to 60 years scientists have been presented with information that will require many of them to tear down and rebuild from the ground up the model they hold of reality. Up until now, most Materialists have taken the easy way by dismissing as anomalies whatever information does not fit their worldview. Enough of these so-called anomalies have now built up that they might be compared to water that has backed up behind a dam. So much is now there that the time has come for the dam to burst.

If you are one of those scientists, it may be helpful to take a look at how we got to the world-view, or model, that's about to be washed downstream.

A Very Old Worldview

There was a time, anthropologists tell us, when humans felt at one with nature. This can still be

129

seen today in primitive cultures. Called pantheism, humans felt they were an integral part of the ecosystem. The divine showed itself in many forms and was present in all things.

But as humans grew more self aware, they began to feel separation. The myth of Adam and Eve recalls the time when humans parted company with the view that they could commune with the divine. They cut the cord by exercising free will.

No longer seeing God in themselves and in others, we humans conjured up gods outside ourselves. In ancient Greece, for example, many gods representing various human qualities were thought to exist. The worldview that evolved in those ancient times had man in the middle between two worlds—a place the Chinese referred to as the Middle Kingdom. The gods lived above the clouds of Mt. Olympus, although they did come to earth now and then, mostly to cause problems for humans.

Below the Middle Kingdom—what caused it to be in the middle—was the underworld, home of the dead, where Hades was in charge and the three-headed dog Cereberus guarded a gate one got to after crossing the River Styx.

Different cultures had different takes on this three-layered universe. Then as now, ideas about

God and gods differed depending on the group one belonged to. The Egyptians had Bal. The Jews had the god of Abraham. The Romans and the Greeks had a pantheon full.

The Idea of One God Evolved and with It a New Worldview

Then along came Jesus of Nazareth and the idea emerged that only one God ruled over creation—although He did have angels and eventually saints who took up some of the positions left vacant by departing Roman and Greek gods.

In 1994 Karen Armstrong published a book, *A History of God: The 4,000-Year Quest of Judaism, Christianity and Islam* (Ballantine Books, 1994), that chronicled history of the emergence of the one-God concept. Because of this idea, the worldview changed somewhat. God and angels replaced the pantheon of gods above the clouds. A fallen angel, Satan, replaced Hades. The place below the ground became hell rather than the underworld—where evildoers went. The good folk would be raised at the end of time on judgment day and be given new, light bodies.

This view held sway for better than a thousand years in Europe and Western Culture but was destined to change again because of a new scientific discovery by Christopher Columbus (1451-1506).

Columbus lived on high ground overlooking a Mediterranean harbor. I have visited the ruin of what is said to be the house where he grew up in Corsica, then part of the city-state of Genoa. In that part of the world there is almost no humidity and the air is very clear. If Columbus had good eyes, he would not have needed a spyglass to see ships climb up over the horizon as they approached the harbor. I've witnessed this myself. Columbus could see the world was round, and he must have decided to prove it by sailing west to get to the spice islands of the East Indies.

Columbus never realized it himself, but he didn't actually get there. Nevertheless, some of Ferdinand Magellan's (1480-1521) crew did, and beyond. Of the 237 men who set out on five ships in 1519, 18 actually completed the circumnavigation of the globe and returned to Spain in 1522.

The newly realized fact that the world was round forced the then commonly held worldview to change. Nevertheless, since people and, most important, Church leaders believed that God had created it, the earth remained at the center of the universe. Now heaven, the dwelling place of God, was seen as being somewhere above the stars. Hell was still beneath the ground, down where it was hot, the place from which molten lava spewed when volcanoes erupted.

The Worldview Gets an Update

It wasn't long before this worldview had to be updated. A fellow named Nicolaus Copernicus (1473–1543) determined the sun was at the center of the solar system. But the Church—the authority back then as science is today—pretty much ignored this concept because it did not go along with accepted canon. This "look the other way" tactic has been used time and again in the twentieth and in the twenty-first century by Materialists.

A century later, along came Galileo Galilei (1564–1642), a man who would not leave well enough alone. Galileo—among other things an astronomer—championed Copernicus's assertion as proven fact. As a result, Galileo started having to watch his back. This was heresy. At that time people were being burned at the stake for less. Indeed, the leaders of the Church told Galileo he'd better recant, and he did. As a result, Galileo got off easy, spending the final years of his life under house arrest on orders of the Inquisition.

But even the Church couldn't keep word from getting out. Gradually, the accepted views of the day began to change.

A Tiny World Comes into View

In 1675, a Dutchman named Antoni van Leeuwenhoek (1632-1723) —an amateur lens grinder and microscope builder—saw for the first time tiny organisms he called "animalcules" living in stagnant water. He also spotted them in scum collected from his teeth. Leeuwenhock didn't know or even speculate that "animalcules" might cause disease. It took until the nineteenth century for that revelation to dawn. At the time, the idea creatures so small they were invisible to the naked eye entered the body to make a person sick and sometimes die would have seemed totally absurd. It was thought demons and the devil caused such things, or that God did it to punish sinners. In 1692 in Salem, Massachusetts, 18 were hanged and one was crushed to death because they were thought to be witches in league with Satan. No wonder after that, and down until today, the idea of Satan and demons and witchcraft was thought to be pure superstition. To believe in such things was to invite witch-hunts and mass hysteria, and nobody wanted that.

The Age of Reason Dawns

Even so, a new day was dawning, a period alternately referred to as "The Age of Enlightenment" and "The Age of Reason." As mentioned earlier, the English philosopher, Thomas Hobbes (1588-1679), had argued that aside from God—the "first cause" who created the material world—nothing existed that is not of the material world. The logic he used was simple. How could it if God created everything? Of course, at that time, all anyone could see was material stuff.

This view was ultimately to lead to the Clockmaker Theory, the idea that God created the universe, wound it up, let it go, and was no longer involved in its operation. Natural laws also had been created that kept going what had been set in motion. Called Deism, many founding fathers, including Thomas Jefferson subscribed to this view.

Hobbes had a big impact on the Age of Enlightenment, which was to pick up steam in the eighteenth century. But the big kahuna was Sir Isaac Newton (1643-1727), an English physicist, mathematician, astronomer, natural philosopher, alchemist, and theologian. Certainly one of the most influential men of all time, his *Philosophiæ Naturalis Principia Mathematica,* published in 1687, is considered to be the groundwork for most of classical mechanics. Newton described universal gravitation and the three laws of motion that dominated the scientific view of the physical universe at least until the advent of quantum mechanics. It seems safe to say Thomas Hobbes's materialistic view of reality coupled with Newton's mechanistic view is the bedrock of most scientific thinking today, except among quantum physicists.

The prevailing worldview that emerged from the Age of Reason was that the universe might be compared to a giant machine. The Sun was at the center of the solar system. The Earth and planets revolved around it. Nothing existed but the material world. What was thought of in the seventeenth century and before as the invisible world of spirit did not exist. Everything that happened had a logical cause. Natural laws governed everything.

Darwin's Theory Takes Hold

In 1859 an Englishman, Charles Darwin (1809-1882), published *On the Origin of Species,* a seminal work in scientific literature and a landmark work in evolutionary biology. Its full title, *On the Origin of Species by Means of Natural Selection, or the Preservation of Favoured Races in the Struggle for Life,* uses the term "races" to mean biological varieties. Darwin's book introduced the theory that populations evolve over the course of generations through a process of natural selection. It presented a body of evidence indicating the diversity of life arose through a branching pattern of evolution and common descent. In other words, God had not created the variety of life on the planet, nor had He created humans. All this had happened through a natural—what might be seen as a mechanical—process. This became accepted as fact among the educated classes.

Darwin's theories reinforced the rationalist idea that the so called supernatural was a figment of human imagination and—not wanting to be burned at the stake, figuratively or literally—most scientists probably wanted to keep it safely buried. Life and its diversity were results of a natural process known as "Survival of the Fittest" coupled with the environment in which a particular species had evolved. Intelligence and mind had evolved as life had evolved and had reached its pinnacle in humans. Mind and intelligence were produced by an organ, the brain, which had come about through evolution. Thought was created by the brain and would later be envisioned as being a result of electrons jumping across synapses. It was contained within the skull. ESP was impossible and so was magic or any other sort of paranormal happening.

A Wedge Between Science and Religion Is Hammered in

With this worldview, a wedge was inserted and hammered in between science, religion and any possibility of things so called supernatural. Hobbes had said nothing existed but the physical. If this were so, where could God possibly reside? What about the heavenly hosts? Thought was contained within the skull so what possible good could prayer do?

A line had been drawn. Educated men and women could not believe in God and prayer or angels or ghosts and demons, which were seen as figments of ignorance and superstition. Many may have had a yearning for God—as humans seem to for the spiritual—but could not rationalize His existence. All were forced to choose between religion and science, though many attempted to straddle the line—as they still do today.

Now, in the first quarter of the twenty-first century, this worldview continues to be the only socially acceptable one in some circles. But there are signs it is beginning to crumble. Hundreds of thousands, perhaps millions, have shifted to a new worldview based on a new branch of science called quantum mechanics and the findings of scientific research that do not fit the materialist-reductionist mold. This new worldview accepts that consciousness is the ground of being of physical reality. I hope this book will play a constructive role in helping more and more people accept this as fact.

Pioneers Leading the Way

Let's look at some have not been afraid to speak out, as well as their ideas and discoveries that conflict with the prevailing nineteenth-twentieth century materialist worldview. The following does not in any way represent an exhaustive list. My apologies to anyone who feels left out, and to anyone who thinks I have overlooked a key figure.

Matter = Energy

In 1905, Albert Einstein (1879-1955), a German-born theoretical physicist, published a paper proving that light behaves both as a wave and as particles. This, as well as Einstein's famous formula, $E = MC^2$, indicates reality and matter are not what they seem. Matter or mass as it is referred to in this formula is equivalent to energy and vice versa.

139

Quantum physicists came along later and expanded on Einstein's work. Niels Henrik David Bohr (1885-1962), a Danish physicist, made fundamental contributions to understanding atomic structure and quantum mechanics, for which he received the Nobel Prize in Physics in 1922. He is quoted as having said, "Everything we call real is made of things that cannot be regarded as real."

Nothing is really solid. Everything is energy—vibrations or waves.

The Collective Unconscious

In 1912 Swiss psychiatrist Carl Jung (1875-1961) published *Wandlungen und Symbole der Libido* (known in English as *The Psychology of the Unconscious)* that postulated a collective unconscious, sometimes known as collective subconscious. According to Jung there is an unconscious mind shared by a society, a people, or all humanity, that is the product of ancestral experience and contains such concepts as the classic archetypes, science, religion, and morality.

ESP and Psycho Kinesis Are Demonstrated to Be Real

As we know from our earlier discussion, in the early 1930s a man named J. B. (Joseph Banks) Rhine moved from Harvard University to Duke to set up a parapsychology laboratory. Rhine not only founded the parapsychology lab at Duke, he also founded the *Journal of Parapsychology* and the Foundation for Research on the Nature of Man. His double blind studies conducted largely between 1930 and 1960 established that ESP exists and is real. Not mentioned in our earlier discussion, they also showed psycho kinesis—mind over matter—is real as well, at least to a statistically significant degree.

His findings were either scoffed at or ignored by the scientific community then as they continue to be today.

Zen Is Introduced to the West

In 1953, Eugen Herrigel (1884-1955), a German philosopher who taught philosophy at Tohoku Imperial University in Sendai, Japan, from 1924-1929 published the book, *Zen and the Art of Archery* (Vintage Books, 1999). This introduced Zen Buddhism to the West and the concept that "All Is One," i.e., everything is connected rather than made up of separate parts. How else could Zen masters shoot arrows while blindfolded and consistently hit the bull's-eyes of targets many yards away?

In 1966 a British philosopher named Alan Watts (1915-1973) published a book called *The Book:*

On the Taboo Against Knowing Who You Are that went into detail about Buddhist thought. Known as an interpreter and popularizer of Asian philosophies for a Western audience, Watts wrote more than 25 books and numerous articles on subjects such as personal identity, the true nature of reality, higher consciousness and the meaning of life. His writings and ideas fueled a new movement which came to be known as "New Age."

Plants Tune into Thoughts

As mentioned earlier, in 1966 a polygraph expert named Cleve Backster (1924-2013) began research that demonstrated living plants tune into the thoughts and intentions of humans as well as other aspects of their environments, thus indicating some sort of hidden mental connection between living things. His findings were ridiculed, but have since been confirmed by other researchers.

Near Death Experiences (NDEs) Are Studied

In 1978 a young man with a B.A., M.A., and Ph.D. from the University of Virginia and an M.D. from Georgia Medical School named Raymond Moody (born 1944) published a book called *Life After*

Life, in which he detailed the experiences of people who had been clinically dead and resuscitated.

The Phenomenon of Grace Is Publicized

Also in 1978, a psychiatrist named M. Scott Peck (1936-2005) published a book that became a huge bestseller called, *The Road Less Travelled: A New Psychology Of Love, Traditional Values And Spiritual Growth.* Among other things, Peck's book dealt with the phenomenon of grace. He said grace was both common and to a certain extent, predictable. He also wrote that, "grace will remain unexplainable within the conceptual framework of conventional science and 'natural law' as we understand it."

Grace is the unseen force that brings the best possible results out of unfortunate events and circumstances, i.e., "every cloud has a silver lining." In Peck's own words, "There is a force, the mechanism of which we do not fully understand, that seems to operate routinely in most people to protect and encourage their physical health even under the most adverse conditions." His book gives specific examples.

Quantum Physics Is Introduced to the Masses

In 1979, Gary Zukav, a former Green Beret during the war in Vietnam, published a book called *The Dancing Wu Li Masters: An Overview of the New Physics.* Targeted for laymen, it explained the basics of quantum physics in everyday language, i.e., without the use of complicated mathematics. Zukav concluded that "the philosophical implication of quantum mechanics is that all of the things in our universe (including us) that appear to exist independently are actually parts of one all-encompassing organic pattern, and that no parts of that pattern are ever really separate from it or from each other."

Earth Is Seen as an Organism

Also in 1979, James Lovelock published a book called *Gaia: A New Look at Life on Earth* that explained his idea that life on earth functions as a single organism. In contrast to the conventional belief that living matter is passive in the face of threats to its existence, the book explored the hypothesis that the earth's living matter—air, ocean, and land surfaces—forms a complex system that has the capacity to keep the Earth a fit place for life. Since

Gaia was first published, many of Jim Lovelock's predictions have come true.

The Spiritual Dimension Is Explored

In the mid 1980s a television series appeared on PBS called *The Power of Myth*, featuring author and Sarah Lawrence College Comparative Religion Professor, Joseph Campbell (1904-1987). These programs made an impact on a significant segment of the public and opened their eyes to the possibility of the existence of what might be termed "a spiritual dimension." This can be summed up using Campbell's own words, "Anyone who has had an experience of mystery knows there is a dimension of the universe that is not that which is available to his senses."

Scientific Studies Demonstrate the Efficacy of Prayer

As previously covered, in July 1988, Dr. Randolph Byrd, a cardiologist, published an article in the Southern Medical Journal about the effects of prayer on cardiac patients. Over a ten-month period, he used a computer to assign 393 patients admitted to the coronary care unit at San Francisco General Hospital either to a group that was prayed for by home prayer groups (192 patients), or to a group that was not prayed for (201). A double blind test, neither the patients, doctors, nor the nurses knew which group a patient was in.

The patients who were remembered in prayer had remarkably, and a statistically significant number of better experiences and outcomes than those who were not prayed for. Also, fewer prayed-for patients died, although the difference between groups in this case was not large enough to be considered statistically significant.

Morphogenetic Fields Are Postulated

In 1994 Rupert Sheldrake, a British biochemist whose theory has already been discussed, published a book called *A New Science of Life.* The editors of the British journal, *Nature,* no doubt being faithful Scientific Materialists, called this book, "the best candidate for burning there has been for many years."

What Researchers Know Can Determine the Outcome

As covered in Chapter Two, Raymond Chiao, a Hong Cong native and quantum physicist then teaching at the University of California at Berkeley, published a paper in 1995 about a series of experiments. The paper, reported upon in a July 1995 issue of *Newsweek* magazine, said that what researchers knew or did not know about certain aspects of each experiment had a predictable determination on their outcomes. In other words, what was in the researchers' minds—i.e. thought—apparently determined the result. In the *Newsweek* article reporting on this, Nobel Prize winning physicist Richard Feynman was quoted as having said this is the "central mystery" of quantum mechanics, that something as intangible as knowledge—in this case, which slit a photon went through—changes something as concrete as a pattern on a screen.

Remote Viewing Is Used by the U.S. Government to Spy

In 2001, F. Holmes ("Skip") Atwater published a book detailing how in 1979 he set up and managed—until his retirement from the Army in 1988—a remote viewing unit of U. S. Army intelligence. The book was entitled, *Captain of My Ship, Master of My Soul: Living With Guidance* (Hampton Roads Publishing Company, 2001).

Prayer Adds Fuel to the Life Force

Also in 2001, as you know, a study published in the September issue of the *Journal of Reproductive*

Medicine showed that prayer was able to double the success rate of in vitro fertilization procedures that lead to pregnancy. The findings revealed that a group of women who had people praying for them had a 50 percent pregnancy rate compared to a 26 percent rate in the group of women who did not have anyone praying for them. In the study—led by Rogerio Lobo, chairman of obstetrics and gynecology at Columbia University's College of Physicians & Surgeons—none of the women undergoing the IVF procedures knew about the prayers on their behalf. Nor did their doctors. In fact, the 199 women were in Cha General Hospital in Seoul, Korea, thousands of miles from those praying for them in the U.S., Canada and Australia. This collaborates with other studies and quantum physics theory that distance is not a factor at the subatomic level of mind.

Research Tells How Best to Pray

An organization exists that has as its purpose the study of what prayer techniques produce the best results. It's called Spindrift and was founded by Christian Science practitioners who have been at this since 1975.

The first question Spindrift researchers sought to answer is, does prayer work? The answer, as we already know, is yes, according to their research. The study provides clues concerning prayer techniques that deliver the best results.

Studies similar to this have been and are being carried out by a consortium of scientists put together by Lynne McTaggart, author of the book published in 2002, *THE FIELD: The Quest for the Secret Force of the Universe*, and her 2008 release, *The INTENTION EXPERIMENT: Using Your Thoughts to Change Your Life* and the World. When she was on my show in early 2008, she described some of these experiments and the terrific success she and her colleagues have had. She said several of these studies were already being prepared for publication.

Mind Is Shown to Create Matter

In 2007, Stephen E. Braude published the book already discussed, *The Gold Leaf Lady and Other Parapsychological Investigations.* The book tells the story of Katie, a woman who demonstrates mind can produce matter—in this case brass: 80% copper and 20% zinc with its huge implications for quantum physics and the origins of the physical universe.

Mediums Can Relate Accurate Information about the Dead

Also in 2008, Julie Beischel, Ph.D., published a paper in *The Journal of Parapsychology* in which she concluded that "certain mediums can report accurate and specific information about the deceased loved ones (termed discarnates) of living people (termed sitters) even without any prior knowledge about the sitters or the discarnates and in the complete absence of any sensory sitter feedback. Moreover, the information reported by these mediums cannot be explained as a result of fraud or 'cold reading' (a set of techniques in which visual and auditory cues from the sitter are used to fabricate 'accurate' readings) on the part of the mediums or bias on the part of the sitters."

A Scientist Will Soon Postulate that God Exists

On 21 April 2020, Stephen Meyer's new book, *The Return of the God Hypothesis: Compelling Scientific Evidence for the Existence of God* (HarperOne, 2020), will be published, and it will present what the publisher calls "groundbreaking scientific evidence of the existence of God, based on breakthroughs in physics, cosmology, and biology." This may be the straw that breaks the back of the Scientific Materialist camel.

Chapter Eight
The Solution

The solution to the "Hard Problem" is that the brain does not create consciousness. Consciousness is ubiquitous, and the brain is the interface between our bodies our personal portion of it.

The evidence suggests that consciousness is the ground of being of physical reality. This being the case, our understanding of reality has come full circle. Long ago, our ancestors thought nature was alive, and they felt at-one with it. Those in primitive cultures today continue to understand and relate to reality this way. Called pantheism, humans believed they were an integral part of the ecosystem and that the divine showed itself in many forms and was present in and expressed itself though all things both alive and inanimate. The evidence that has been revealed in this book indicates to me that this view is accurate. How about you? Having read almost this entire book, do you think it's likely that consciousness underlies, supports and informs the physical world? If you don't think so, please let me know why at www.shmartin.com.

And what about our personal consciousness? Where did it come from?

Perhaps, as we grow from infants into adults, our brains shape a tiny bit of the Infinite Mind into our personal, "I am" consciousness. However, since we know from UVA's research that reincarnation can and does happen, an alternative possibility is that our personal, "I am" consciousness may have become differentiated from the whole billions of years ago in the primordial sea. If so, it has evolved over the eons during many incarnations as our bodies have evolved, birth after birth. Of those two possibilities, the latter is the one I favor. But whatever the case may be, after thousands of years, and a number of different ways of looking at reality, we have arrived back where we began, although on a higher level of understanding. It will probably take at least a few generations for this new worldview to be widely held, but I believe it is only a matter of time before it is. This is because it makes sense, or, as Arthur Schopenhauer seemed to indicate in his quip about the three stages of the acceptance of truth: for those willing to see, it is self-evident. It is certainly in line with what we know about quantum mechanics. As was mentioned earlier, Gary

Zukav wrote in his book on that subject, *The Dancing Wu Li Masters:*

> *. . . the philosophical implication of quantum mechanics is that all of the things in our universe (including us) that appear to exist independently are actually parts of one all-encompassing organic pattern, and that no parts of that pattern are ever really separate from it or from each other.*

So it appears that the entire universe indeed is one huge interconnected whole, and we and other living things are concentrations of consciousness that exist within the whole. Earlier I mentioned that I once had what might be described as a "mystical experience." When it happened I understood intuitively that "all-is-one," which apparently is a fairly common experience. In his book, *The Rebirth of Nature, The Greening of Science and God,* Rupert Sheldrake quotes a woman, an art teacher, recounting an experience she had while she was walking on the Pangbourne Moors at the age of five. She put into words what I felt when I had my personal experience of mystery:

Suddenly I seemed to see the mist as a shimmering gossamer tissue and the harebells, appearing here and there, seemed to shine with a brilliant fire. Somehow I understood that this was the living tissue of life itself, in which all that we call consciousness is embedded, appearing here and there as a shining focus of energy in the more diffused whole. In that moment I knew that I had my special place, as had all other things, animate and so-called inanimate, and that we were all part of this universal tissue which was both fragile yet immensely strong, and utterly good and beneficent.

There, you have it. We are all part of an "all-encompassing organic pattern," the implications of which are huge. I explore some of them in other books, but will leave you now with one that you may find to be life-changing:

You are consciousness itself, a unique expression of the whole, and like the whole, you are eternal and only a small way toward your ultimate destination.

Let that sink in and see if it doesn't change how you view your life and the world you live in.

# # #

Stephen Hawley Martin is an author, ghostwriter, and publisher. You can learn more about him and get in touch with him through his website:

www.shmartin.com

If you found this book interesting, other books by Stephen you should know about are displayed on the pages that follow.

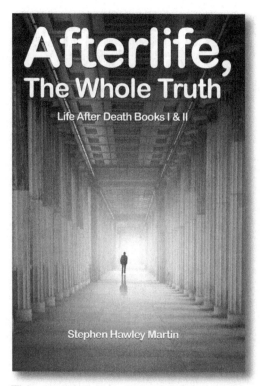

This two-book volume contains the bestselling title, *Life After Death, Powerful Evidence You Will Never Die* and the sequel, *Heaven, Hell & You.* As one reviewer, a medical doctor, wrote: "Extraordinary findings . . . will keep readers on the edge of their seats as they burn through this well written book's pages."

Kindle: ASIN: B07J46QQW8
Paperback: ISBN-10: 1727782038

The True Holy Grail

The Secret You Can Use To Create the Life You Want

Stephen Hawley Martin

Some searching for the Grail thought it was the cup Jesus used at the Last Supper. Others, a dish, or a stone. All believed it had powers to create happiness and wealth. This book explains that the True Holy Grail is secret knowledge that enabled Jesus to work miracles. This book tells that secret.

Kindle: ASIN: B07MYQ8N8V
Paperback: ISBN-10: 1794500715

Stephen Hawley Martin

A WITCH IN THE FAMILY

a witch in the family

The Salem Witch Trials Re-examined
In Light of New Evidence

Second Edition

Nineteen were hanged, including the author's seven-times-great grandmother, and one was crushed to death. Were their accusers lying as indicated in Arthur Miller's play, "The Crucible?" Maybe not. This book is a riveting account of a real-life murder mystery. If you'd like to know what really happened in 1692 Salem, don't miss this book.

Kindle: ASIN: B06XY8WWJ6
Paperback: ISBN-10: 1093653701

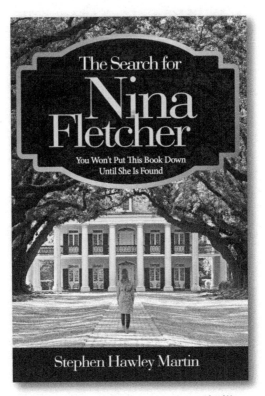

The Search for **Nina Fletcher**

You Won't Put This Book Down
Until She Is Found

Stephen Hawley Martin

In this romantic suspense thriller, Rebecca wants to save the beautiful plantation home where she grew up, but to do so she must find her mother. If only she could remember what happened in the basement of the old house in Baltimore long ago. She must find out what happened there, she must!

Kindle: ASIN: B01J6MQZXS
Paperback: ISBN-10: 1535580879

Death in Advertising

FICTION FIRST PRIZE WINNER WRITER'S DIGEST

Stephen Hawley Martin

This whodunit set in an ad agency won First Prize for Fiction from *Writer's Digest* magazine. According to Mike Chapman, Editor-in-Chief of *ADWEEK* magazine, this novel is "A thrilling and evocative read. Masterful attention to detail brings the ad agency world to life and delivers a gripping whodunit." Get ready. You won't be able to put it down.

Kindle: ASIN: B00UIGGKUA
Paperback: ISBN-10: 1511662921

Based on a Catastrophic Historical Event

FICTION FIRST PRIZE
WINNER
WRITER'S DIGEST

The Mt Pelée Redemption

**A Thriller by
Stephen Hawley Martin**

Romantic suspense at its best, this fast-paced novel won First Prize for Fiction from *Writer's Digest* and First Place for Visionary Fiction from *Independent Publisher* for good reason: It's very hard to put down. You'll be riveted as Claire flies to the island of Martinique to solve a mystery and soon realizes she's being stalked by a drug lord.

Kindle: ASIN: B00UVK8XM6
Paperback: ISBN-10: 1511675373

ESP

How I Developed My Sixth Sense
And So Can You

Stephen Hawley Martin

All the knowledge of the universe resides within you because at a deep level all minds, past and present, are connected. Everything that has ever happened, every thought, every idea is there. The trick is to draw out information when you need it. In this book Stephen explains how he learned to do so and how you can, too.

Kindle: ASIN: B07HHFFWP8
Paperback: ISBN-10: 1723835250

Made in the USA
Monee, IL
21 December 2019